Ben Stacy Jerrik (Ed.)

MassTransit Enterprise

Ben Stacy Jerrik (Ed.)

MassTransit Enterprise

Managed file transfer, GroupLogic, Macintosh

Part Press

Imprint

Publisher:
Part Press is a trademark of
International Book Market Service Ltd., 17 Rue Meldrum, Beau Bassin, 1713-01 Mauritius
Email: info@bookmarketservice.com
Website: www.bookmarketservice.com

Published in 2012

Printed in: U.S.A., U.K., Germany. This book was not produced in Mauritius.

ISBN: 978-613-9-39669-6

Contents

Articles

References

MassTransit_Enterprise

Developer(s)	Group Logic, Inc.
Stable release	7.2.3 / June 2011
Operating system	Windows and Macintosh
Type	Managed File Transfer
License	Proprietary
Website	[1]

MassTransit Enterprise, a Managed File Transfer server from Group Logic, Inc, (Arlington, Virginia, USA), runs on Windows and Macintosh operating systems and provides a range of managed file transfer functions.

MassTransit was originally released as Adobe Virtual Network in 1995.

MassTransit Capabilities

MassTransit supports a number of TCP/IP based file transfer protocols including HTTP, HTTPS, FTP, SFTP, an optimized proprietary protocol MTAP, a secure version of MTAP as well as a UDP Data Transport version of MTAP based on User Datagram Protocol (UDP).

MassTransit server (masstransit.exe) requires Windows and supports Windows Server 2008 (R2, 32-bit & 64-bit) and Windows 2003 Server. Active Directory users and groups control access to the transfer and reporting capabilities.

MassTransit Release History

Latest Releases [2]

- MassTransit 7.2.3 June 2011
- MassTransit 7.0.0 1 July 2010
- MassTransit 6.1.1 December 2009
- MassTransit 6.1 November 2009
- MassTransit 6.0.1x04 December 2008
- MassTransit 6.0 for Mac OS X and Windows in November 2008.
- MassTransit 5.1.2 as hot fixes for Mac OS X and Windows from March to July 2008.
- MassTransit 5.1.1 as hot fixes for Mac OS X and Windows in February 2008.
- MassTransit 5.1 9/10/2007 for Mac OS X and Windows.
- MassTransit 5.0.2 as hot fixes for Mac OS X and Windows from January to August 2007.
- MassTransit 5.0.1 November 2006.
- MassTransit 5.0
- MassTransit 4.5.1 as a series of hot fixes from April 2004 through March 2006.
- MassTransit 1.0 released as Adobe Virtual Network 1.0 April 1996

References

- Mac Observer Interviews Group Logic CEO [3]
- BusinessWeek Private Company Information: Group Logic, Inc. [4]
- Key Issues for Managed File Transfer, 2009 19 February 2009 | ID:G00165299 [5]
- Moving Beyond MFT to File Services 29 May 2009 | ID:G00168386 [6]
- Data Encryption Not Enough to Prevent FTP Credential Theft 6 July 2009 | ID:G00169584 [7]
- Hype Cycle for Data and Application Security, 2009 17 July 2009 | ID:G00168605 [8]

External links

- Company Website [9]
- Enterprise Desktop Alliance Website [10]
- Apple solutions guide [11]
- Microsoft solutions directory [12]

References

[1] http://www.grouplogic.com/products/MassTransit/
[2] http://support.grouplogic.com/?page_id=70
[3] http://www.macobserver.com/tmo/article/tmo_interviews_group_logic_wwdc_is_eye_opener_for_executives/
[4] http://investing.businessweek.com/research/stocks/private/snapshot.asp?privcapId=29174
[5] http://www.gartner.com/DisplayDocument?doc_cd=165299
[6] http://www.gartner.com/DisplayDocument?id=998412
[7] http://www.gartner.com/DisplayDocument?ref=g_search&id=1059118&subref=simplesearch
[8] http://www.gartner.com/DisplayDocument?doc_cd=168605
[9] http://www.grouplogic.com/
[10] http://www.enterprisedesktopalliance.com/
[11] http://guide.apple.com/action.lasso?-database=macosguide&-layout=cgi_detail&-response=/ussearch/detail.html&prodkey=41899&-search
[12] https://solutionfinder.microsoft.com/Solutions/SolutionDetailsView.aspx?solutionid=75b1c23c811e444c8307197a9ac8ed47&Partnerid=ea539ee6-265f-4e95-bb70-7eca94f80f06

Managed_file_transfer

Managed file transfer (MFT) refers to software solutions that facilitate the secure transfer of data, in flight and at rest, from one computer to another through a network (e.g., the Internet). MFT solutions are often built to support the FTP network protocol. However, the term specifically describes solutions that remedy the disadvantages associated with FTP.

Typically, MFT offers a higher level of security and control than FTP. Features include reporting (e.g., notification of successful file transfers), non-repudiation, auditability, global visibility, automation of file transfer-related activities and processes, end-to-end security, and performance metrics/monitoring.

MFT applications are available as both licensed software packages and SaaS solutions. Some are specially designed for enterprise use while others are for sale to individual consumers. A few enterprise-focused SaaS MFT providers also manage the additions of new trading partners, which can free up a lot of IT resources.

MFT Solutions are characterized by having all or most of the following features:

- Support multiple file transfer protocols including FTP/S, OFTP, SFTP, SCP, AS2, and HTTP/S.
- Securely transfer files over public and private networks using encrypted file transfer protocols.
- Securely store files using multiple data encryption methods
- Automate file transfer processes between trading partners and exchanges including detection and handling of failed file transfers.
- Authenticate users against existing user repositories such as LDAP and Active Directory
- Integrate to existing applications using documented API
- Generate detailed reports on user and file transfer activity.

Background

From its inception, FTP has made moving large volumes of bulk data between any two entities—including file servers, applications, and trading partners—possible. However, FTP (and other communication protocols such as HTTP and SMTP) do not, on their own, provide a way to secure or manage the payload or the transmission. Yet, regardless of the lack of security and management capabilities, many companies have continued to transport large batches of structured and unstructured data using these protocols.

This practice is changing, however. According to Gartner Research, "Organizations often use MFT solutions to replace FTP. This is due to increased focus on compliance, privacy regulations and corporate transparency — which demand increased auditing, management, security and process.[1]

Different flavors

While Managed File Transfer always recovers the same features -reporting (e.g., notification of successful file transfers), non-repudiation, auditability, global visibility, automation of file transfer-related activities and processes, end-to-end security, and performance metrics/monitoring- the way it is used has a major impact on the nature of the appropriate solution. Today analysts [2] and experts agree on 6 different usage patterns for MFT:

1. Ad Hoc
2. Accelerated Transfer or "Extreme" Transfer
3. A2A (Administrative-2-Administration)
4. B2B
5. Cloud
6. SOA

All are defined by their own markers and require different solutions.

Future

As more and more companies decide that FTP is inappropriate for secure file transfer[3] , Managed File Transfer is beginning to be perceived as a necessity, not an option.[4]

Challenges for MFT systems include:

- adaptability to the wide variety of business and technical requirements for file transfers within a wide variety of organizations
- scalability: the ability for the capacity of MFT systems to expand along with an organization's growth
- SAAS model adaptability

References

[1] "FTP Replacement: Where MFT Makes Sense and Why You Should Care" (http://www.gartner.com/DisplayDocument?doc_cd=208765). 08-NOV-2010. .

[2] "What MFT Is, and How It Applies to You" (http://www.gartner.com/DisplayDocument?doc_cd=214111). 14-JUN-2011. .

[3] "FTP Replacement: Where MFT Makes Sense and Why You Should Care (45$ from Gartner)" (http://www.gartner.com/DisplayDocument?doc_cd=208765). 08-NOV-2010. .

[4] Woodie, Alex (2 September 2009). "Managed File Transfer: A New Product Category That's Here to Stay" (http://www.itjungle.com/fhs/fhs090109-story01.html). IT Jungle-Four Hundred Stuff. .

GroupLogic

GroupLogic, Inc. Logo - 5/2011

Type	Private
Industry	Technology
Founded	August 1, 1988
Headquarters	Arlington, VA
Key people	Chris Broderick (CEO) T. Reid Lewis (President) David Stokley (CFO) Derick Naef (CTO) Andy Lewis (SVP) Anders Lofgren (VP, Product Management) Summer Benish (VP, Sales) Todd While (VP, Mobility Solutions)
Employees	40
Website	www.grouplogic.com [1]

GroupLogic, Inc. [1], founded in 1988 and headquartered in Arlington, VA, USA, is an enterprise software company that develops, sells and supports software for moving and storing data including launched activEcho, mobilEcho, ArchiveConnect [2], MassTransit and ExtremeZ-IP. GroupLogic's products are used by Information Technology(IT) organizations to allow employees to access and manage corporate files regardless of the type of computing platform - such as Apple or PC - the employee is using to access the network.

On May 23, 2011 GroupLogic launched mobilEcho, a mobile file management system for iPad. mobilEcho (mE)is installed on corporate file servers by IT departments and installed on iPads and tablet computers by users; GroupLogic says mobilEcho allows for the storage and secure transfer of files back and forth directly between file servers and iPads.[3]

On March 13, 2012 GroupLogic launched activEcho, an enterprise file sharing software solution with clients for Windows, Mac OS X, iPhone, iPad and popular web browsers and deployment options on Windows servers on premise or in data centers or private clouds.

Company history

1988 Founded

1990 Shipped Aspects realtime collaborative editing product

1991 MacUser Magazine recognized Aspects with the Editors Choice Award

1992 Shipped Whiteboard for Windows and Macintosh

1992 Shipped ImageXpo for Windows and Macintosh

1995 Shipped Adobe Virtual Network (now MassTransit Enterprise)

1999 Shipped ExtremeZ-IP

2008 Co-founded Enterprise Desktop Alliance

2009 Shipped ArchiveConnect [4]

2010 Shipped ShadowConnect [5]

2011 Launched mobilEcho [6]

2012 Launched activEcho [7]

References

[8] [9]

[1] http://www.grouplogic.com
[2] http://www.grouplogic.com/enterprise-file-sharing/mac-file-archiving-system/
[3] "GroupLogic Brings Enterprise-Friendly File Management Platform To The iPad" (http://techcrunch.com/2011/05/22/grouplogic-brings-enterprise-friendly-file-management-platform-to-the-ipad/). . Retrieved 22-May-2011.
[4] http://www.grouplogic.com/products/ArchiveConnect/
[5] http://www.grouplogic.com/products/ShadowConnect/
[6] http://www.grouplogic.com/enterprise-file-sharing/ipad-file-system/
[7] http://www.grouplogic.com/enterprise-file-sharing/file-sync/
[8] ESG Enterprise Strategy Group "GroupLogic Dives Into the Online File Storage Pool" (http://www.itdependsblog.com/2012/03/13/grouplogic-dives-into-the-online-file-storage-pool/). ESG Enterprise Strategy Group. Retrieved 13-March-2012.
[9] ReadWriteWeb "GroupLogic Has New Way To Securely Synch Your Files With ActivEcho" (http://www.readwriteweb.com/enterprise/2012/03/grouplogic-has-new-way-to-secu.php). ReadWriteWeb. Retrieved 13-March-2012.

- Slate Group Logic: What can a tiny software firm show us about the future of American exports 16-Nov-2010 (http://www.slate.com/id/2274793/)
- Mac Observer Interviews Group Logic CEO (http://www.macobserver.com/tmo/article/tmo_interviews_group_logic_wwdc_is_eye_opener_for_executives/)
- BusinessWeek Private Company Information: Group Logic, Inc. (http://investing.businessweek.com/research/stocks/private/snapshot.asp?privcapId=29174)
- Key Issues for Managed File Transfer, 2009 19 February 2009 | ID:G00165299 (http://www.gartner.com/DisplayDocument?doc_cd=165299)
- Moving Beyond MFT to File Services 29 May 2009 | ID:G00168386 (http://www.gartner.com/DisplayDocument?id=998412)
- Data Encryption Not Enough to Prevent FTP Credential Theft 6 July 2009 | ID:G00169584 (http://www.gartner.com/DisplayDocument?ref=g_search&id=1059118&subref=simplesearch)
- Hype Cycle for Data and Application Security, 2009 17 July 2009 | ID:G00168605 (http://www.gartner.com/DisplayDocument?doc_cd=168605)

External links

- Company Website (http://www.grouplogic.com/)
- Company YouTube Channel (http://www.youtube.com/user/GroupLogic)
- Enterprise Device Alliance Website (http://www.enterprisedesktopalliance.com/)
- Microsoft solutions directory (http://pinpoint.microsoft.com/en-US/PartnerDetails.aspx?PartnerId=4295786962)

Macintosh

The **Macintosh** (/ˈmækɪntɒʃ/ ***MAK****-in-tosh*),[1] or **Mac**, is a series of personal computers (PC) designed, developed, and marketed by Apple Inc. The first Macintosh was introduced by Apple's then-chairman Steve Jobs on January 24, 1984; it was the first commercially successful personal computer to feature a mouse and a graphical user interface rather than a command-line interface.[2] The company continued to have success through the second half of the 1980s, first primarily because of the sales of the Apple II series remained strong even after the introduction of the Macintosh, only to see it dissipate in the 1990s as the personal computer market shifted toward the "Wintel" platform: IBM PC compatible machines running MS-DOS and Microsoft Windows.[3]

The original Macintosh, the first commercially successful personal computer to use a graphical user interface, rather than a command line.

In 1998, Apple consolidated its multiple consumer-level desktop models into the iMac all-in-one. This proved to be a sales success and saw the Macintosh brand revitalized. Current Mac systems are mainly targeted at the home, education, and creative professional markets. These include the descendants of the original iMac and the entry-level Mac mini desktop models, the Mac Pro tower graphics workstation, the MacBook, MacBook Air and MacBook Pro laptops. The Xserve server was discontinued January 31, 2011.[4]

An iMac computer from October 2009, a modern all-in-one Macintosh.

Production of the Mac is based on a vertical integration model in that Apple facilitates all aspects of its hardware and creates its own operating system that is pre-installed on all Mac computers. This is in contrast to most IBM PC compatibles, where multiple sellers create and integrate hardware intended to run another company's operating software. Apple exclusively produces Mac hardware, choosing internal systems, designs, and prices. Apple does use third party components, however, such as graphics subsystems from nVidia and ATi. Current Mac CPUs use Intel's X86-64 architecture; the earliest models (1984–1994) used Motorola's 68k and models from 1994–2006 used the AIM alliance's PowerPC. Apple also develops the operating system for the Mac, currently Mac OS X version 10.7 "Lion". The modern Mac, like other personal computers, is capable of running alternative operating systems such as Linux, FreeBSD, and, in the case of Intel-based Macs, Microsoft Windows. However, Apple does not license Mac OS X for use on non-Apple computers.

History

1979 to 1984: Development

The Macintosh project started in the late 1970s with Jef Raskin, an Apple employee who envisioned an easy-to-use, low-cost computer for the average consumer. He wanted to name the computer after his favorite type of apple, the McIntosh,[5] but the name had to be changed for legal reasons as it was too close, phonetically, to that of the McIntosh audio equipment manufacturer. Steve Jobs requested a release of the name so that Apple could use it, but was denied, forcing Apple to eventually buy the rights to use the name.[6] Raskin was authorized to start hiring for the project in September 1979,[7] and he began to look for an engineer who could put together a prototype. Bill Atkinson, a member of Apple's Lisa team (which was developing a similar but higher-end computer), introduced him to Burrell Smith, a service technician who had been hired earlier that year. Over the years, Raskin assembled a large development team that designed and built the original Macintosh hardware and software; besides Raskin, Atkinson and Smith, the team included George Crow,[8] Chris Espinosa, Joanna Hoffman, Bruce Horn, Susan Kare, Andy Hertzfeld, Guy Kawasaki, Daniel Kottke,[9] and Jerry Manock.[10] [11]

Smith's first Macintosh board was built to Raskin's design specifications: it had 64 kilobytes (kB) of RAM, used the Motorola 6809E microprocessor, and was capable of supporting a 256×256 pixel black-and-white bitmap display. Bud Tribble, a Macintosh programmer, was interested in running the Lisa's graphical programs on the Macintosh, and asked Smith whether he could incorporate the Lisa's Motorola 68000 microprocessor into the Mac while still keeping the production cost down. By December 1980, Smith had succeeded in designing a board that not only used the 68000, but increased its speed from 5 MHz to 8 MHz; this board also had the capacity to support a 384×256 pixel display. Smith's design used fewer RAM chips than the Lisa, which made production of the board significantly more cost-efficient. The final Mac design was self-contained and had the complete QuickDraw picture language and interpreter in 64 kB of ROM – far more than most other computers; it had 128 kB of RAM, in the form of sixteen 64 kilobit (kb) RAM chips soldered to the logicboard. Though there were no memory slots, its RAM was expandable to 512 kB by means of soldering sixteen IC sockets to accept 256 kb RAM chips in place of the factory-installed chips. The final product's screen was a 9-inch, 512x342 pixel monochrome display, exceeding the prototypes.[12]

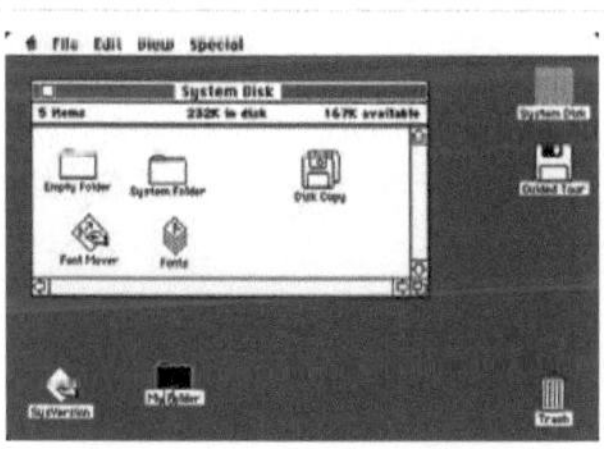

The original 1984 Mac OS desktop featured a radically new graphical user interface. Users communicated with the computer not through abstract textual commands but rather using a metaphorical desktop that included icons of real life items with which the user was already familiar.

The design caught the attention of Steve Jobs, co-founder of Apple. Realizing that the Macintosh was more marketable than the Lisa, he began to focus his attention on the project. Raskin finally left the Macintosh project in 1981 over a personality conflict with Jobs, and team member Andy Hertzfeld said that the final Macintosh design is closer to Jobs' ideas than Raskin's.[7] After hearing of the pioneering GUI technology being developed at Xerox PARC, Jobs had negotiated a visit to see the Xerox Alto computer and Smalltalk development tools in exchange for Apple stock options. The Lisa and Macintosh user interfaces were influenced by technology seen at Xerox PARC and were combined with the Macintosh group's own ideas.[13] Jobs also commissioned industrial designer Hartmut Esslinger to work on the Macintosh line, resulting in the "Snow White" design language; although it came too late for the earliest Macs, it was implemented in most other mid- to late-1980s Apple computers.[14] However, Jobs' leadership at the Macintosh project did not last; after an internal power struggle with new CEO John Sculley, Jobs resigned from Apple in 1985,[15] went on to found NeXT, another computer company,[16] and did not return until 1997 when Apple acquired NeXT.[17]

The first Macintosh computer was manufactured at an Apple plant in Fremont, California.[18]

1984: Introduction

The Macintosh 128K was announced to the press in October 1983, followed by an 18-page brochure included with various magazines in December.[19] The Macintosh was introduced by the now famous US$1.5 million Ridley Scott television commercial, "1984".[20] The commercial most notably aired during the third quarter of Super Bowl XVIII on January 22, 1984 and is now considered a "watershed event"[21] and a "masterpiece."[22] "1984" used an unnamed heroine to represent the coming of the Macintosh (indicated by a Picasso-style picture of Apple's Macintosh computer on her white tank top) as a means of saving humanity from the "conformity" of IBM's attempts to dominate the computer industry. The ad alludes to George Orwell's novel, *Nineteen Eighty-Four,* which described a dystopian future ruled by a televised "Big Brother."[23] [24]

This television commercial, first aired during Super Bowl XVIII, launched the original Macintosh.

Two days after the 1984 ad aired, the Macintosh went on sale. It came bundled with two applications designed to show off its interface: MacWrite and MacPaint. It was first demonstrated by Steve Jobs in the first of his famous Mac Keynote speeches, and though the Mac garnered an immediate, enthusiastic following, some labeled it a mere "toy."[25] Because the operating system was designed largely around the GUI, existing text-mode and command-driven applications had to be redesigned and the programming code rewritten. This was a time-consuming task that many software developers chose not to undertake, and could be regarded as a reason for an initial lack of software for the new system. In April 1984 Microsoft's MultiPlan migrated over from MS-DOS, with Microsoft Word following in January 1985.[26] In 1985, Lotus Software introduced Lotus Jazz for the Macintosh platform after the success of Lotus 1-2-3 for the IBM PC, although it was largely a flop.[27] Apple introduced Macintosh Office the same year with the lemmings ad. Infamous for insulting its own potential customers, the ad was not successful.[28]

For a special post-election edition of *Newsweek* in November 1984, Apple spent more than US$2.5 million to buy all 39 of the advertising pages in the issue.[29] Apple also ran a "Test Drive a Macintosh" promotion, in which potential buyers with a credit card could take home a Macintosh for 24 hours and return it to a dealer afterwards. While 200,000 people participated, dealers disliked the promotion, the supply of computers was insufficient for demand, and many were returned in such a bad shape that they could no longer be sold. This marketing campaign caused CEO John Sculley to raise the price from US$1,995 to US$2,495 (adjusting for inflation, about $5,200 in 2010).[28] [30]

The Apple Macintosh Plus at the Design Museum in Gothenburg, Sweden.

1985 to 1989: Desktop publishing era

In 1985, the combination of the Mac, Apple's LaserWriter printer, and Mac-specific software like Boston Software's MacPublisher and Aldus PageMaker enabled users to design, preview, and print page layouts complete with text and graphics—an activity to become known as desktop publishing. Initially, desktop publishing was unique to the Macintosh, but eventually became available for other platforms as well.[31] Later, applications such as Macromedia FreeHand, QuarkXPress, Adobe Photoshop, and Adobe Illustrator strengthened the Mac's position as a graphics computer and helped to expand the emerging desktop publishing market.

The limitations of the first Mac soon became clear: it had very little memory, even compared with other personal computers in 1984, and could not be expanded easily; and it lacked a hard disk drive or the means to attach one easily. Many small companies sprang up to address the memory issue, by upgrading the 128k Mac to 512 kB, by removing the computer's 16 memory chips and replacing them with larger-capacity chips, a tedious operation that was not always successful. In October 1985, Apple increased the Mac's memory to 512 kB,[32] and offered an upgrade for 128k Macs that involved replacing the logic board. In an attempt to improve connectivity, Apple released the Macintosh Plus on January 10, 1986 for US$2,600. It offered one megabyte of RAM, easily expandable to four by the use of socketed RAM boards, and a SCSI parallel interface, allowing up to seven peripherals—such as hard drives and scanners—to be attached to the machine. Its floppy drive was increased to an 800 kB capacity. The Mac Plus was an immediate success and remained in production, unchanged, until October 15, 1990; on sale for just over four years and ten months, it was the longest-lived Macintosh in Apple's history.[33]

The Macintosh II, one of the first expandable Macintosh models.

Updated Motorola CPUs made a faster machine possible, and in 1987 Apple took advantage of the new Motorola technology and introduced the Macintosh II, which used a 16 MHz Motorola 68020 processor.[34] The primary improvement in the Macintosh II was Color QuickDraw in ROM, a color version of the graphics language which was the heart of the machine. Among the many innovations in Color QuickDraw were an ability to handle any display size, any color depth, and multiple monitors.

The Macintosh II marked the start of a new direction for the Macintosh, as now for the first time it had an open architecture with several NuBus expansion slots, support for color graphics and external monitors, and a modular design similar to that of the IBM PC. It had an internal hard drive and a power supply with a fan, which was initially fairly loud.[35] One third-party developer sold a device to regulate fan speed based on a heat sensor, but it voided the warranty.[36] Later Macintosh computers had quieter power supplies and hard drives.

In September 1986, Apple introduced the Macintosh Programmer's Workshop, or MPW that allowed software developers to create software for Macintosh on Macintosh, rather than cross compiling from a Lisa. In August 1987, Apple unveiled HyperCard, and introduced MultiFinder, which added cooperative multitasking to the Macintosh. Apple began bundling both with every Macintosh.

The Macintosh SE was released at the same time as the Macintosh II, as the first compact Mac with a 20 MB internal hard drive and one expansion slot.[37] The SE's expansion slot was located inside the case along with the CRT, potentially exposing an upgrader to high voltage. For this reason Apple recommended users bring their SE to an authorized Apple dealer to have upgrades performed.[38] The SE also updated Jerry Manock and Terry Oyama's original design and shared the Macintosh II's *Snow White* design language, as well as the new Apple Desktop Bus (ADB) mouse and keyboard that had first appeared on the Apple IIGS some months earlier.

In 1987, Apple spun off its software business as Claris. It was given the code and rights to several applications that had been written within Apple, notably MacWrite, MacPaint, and MacProject. In the late 1980s, Claris released a number of revamped software titles; the result was the "Pro" series, including MacDraw Pro, MacWrite Pro, and FileMaker Pro. To provide a complete office suite, Claris purchased the rights to the Informix Wingz spreadsheet on the Mac, renaming it Claris Resolve, and added the new presentation software Claris Impact. By the early 1990s, Claris applications were shipping with the majority of consumer-level Macintoshes and were extremely popular. In 1991, Claris released ClarisWorks, which soon became their second best-selling application. When Claris was reincorporated back into Apple in 1998, ClarisWorks was renamed AppleWorks beginning with version 5.0.[39]

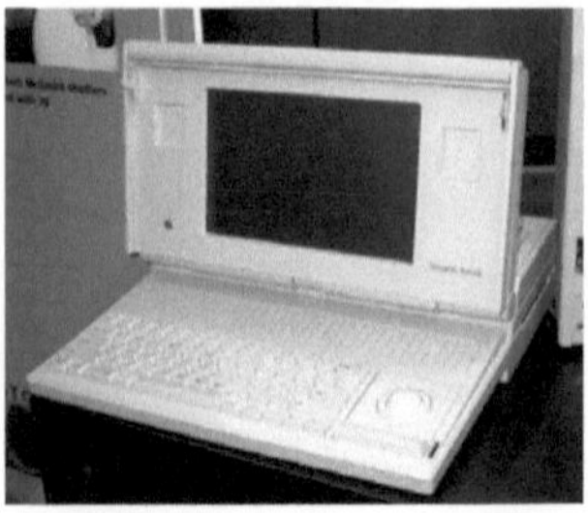
The Macintosh Portable was Apple's first battery-powered Macintosh. It was available from 1989 to 1991 and could run System 6 and System 7.

In 1988, Apple sued Microsoft and Hewlett-Packard on the grounds that they infringed Apple's copyrighted GUI, citing (among other things) the use of rectangular, overlapping, and resizable windows. After four years, the case was decided against Apple, as were later appeals. Apple's actions were criticized by some in the software community, including the Free Software Foundation (FSF), who felt Apple was trying to monopolize on GUIs in general, and boycotted GNU software for the Macintosh platform for seven years.[40] [41]

With the new Motorola 68030 processor came the Macintosh IIx in 1988, which had benefited from internal improvements, including an on-board MMU.[42] It was followed in 1989 by a more compact version with fewer slots (the Macintosh IIcx)[43] and a version of the Mac SE powered by the 16 MHz 68030, the Macintosh SE/30.[44] Later that year, the Macintosh IIci, running at 25 MHz, was the first Mac to be "32-bit clean," allowing it to natively support more than 8 MB of RAM,[45] unlike its predecessors, which had "32-bit dirty" ROMs (8 of the 32 bits available for addressing were used for OS-level flags). System 7 was the first Macintosh operating system to support 32-bit addressing.[46] Apple also introduced the Macintosh Portable, a 16 MHz 68000 machine with an active matrix flat panel display that was backlit on some models.[47] The following year the Macintosh IIfx, starting at US$9,900, was unveiled. Apart from its fast 40 MHz 68030 processor, it had significant internal architectural improvements, including faster memory and two Apple II CPUs dedicated to I/O processing.[48]

1990 to 1998: Growth and decline

The Macintosh Classic, Apple's early 1990s budget model.

Microsoft Windows 3.0, which began to approach the Macintosh operating system in both performance and feature set, was released in May 1990 and was a less expensive alternative to the Macintosh platform. Apple's response was to introduce a range of relatively inexpensive Macs in October 1990. The Macintosh Classic, essentially a less expensive version of the Macintosh Plus, was the least expensive Mac until early 2001.[49] The 68020-powered Macintosh LC, in its distinctive "pizza box" case, offered color graphics and was accompanied by a new, low-cost 512 × 384 pixel monitor.[50] The Macintosh IIsi was essentially a 20 MHz IIci with only one expansion slot.[51] All three machines sold well,[52] although Apple's profit margin was considerably lower than on earlier machines.[49]

The PowerBook 100 (shown here), 140 and 170 introduced a line of professional laptop Macs. They pioneered notebook ergonomics by placing the keyboard behind a palm rest.

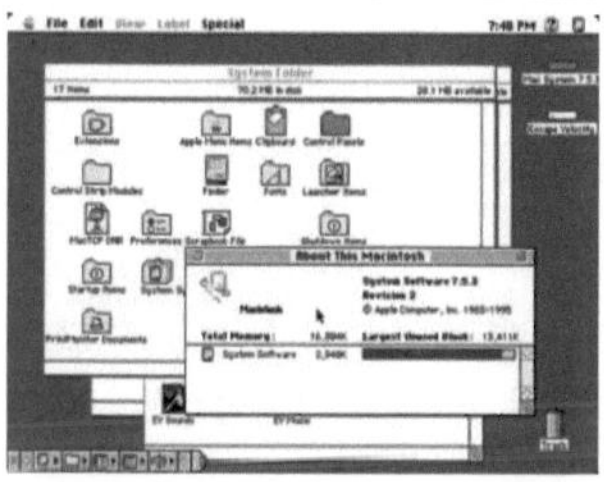

System 7 was the first major upgrade of the Macintosh operating system.

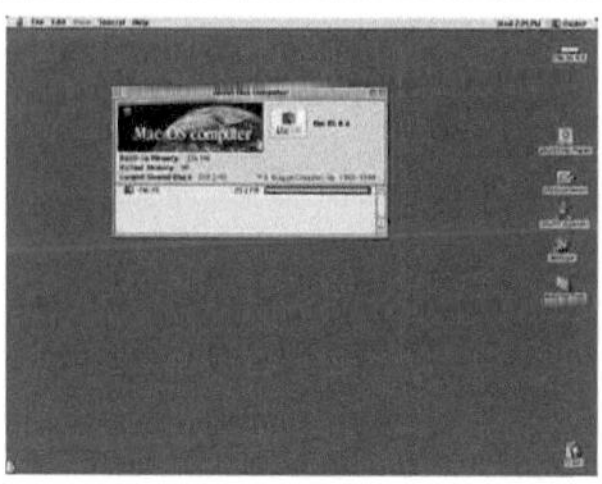

OS 8 was the second major upgrade of the Mac OS. OS 8.6 shown.

Apple improved Macintosh computers by introducing models equipped with newly available processors from the 68k lineup. The Macintosh Classic II[53] and Macintosh LC II, which used a 16 MHz 68030 CPU,[54] were joined in 1991 by the Macintosh Quadra 700[55] and 900,[56] the first Macs to employ the faster Motorola 68040 processor. In 1994, Apple abandoned Motorola CPUs for the RISC PowerPC architecture developed by the AIM alliance of Apple Computer, IBM, and Motorola.[57] The Power Macintosh line, the first to use the new chips, proved to be highly successful, with over a million PowerPC units sold in nine months.[58]

Apple replaced the Macintosh Portable in 1991 with the first of the PowerBook line: the PowerBook 100, a miniaturized Portable; the 16 MHz 68030 PowerBook 140; and the 25 MHz 68030 PowerBook 170.[59] They were the first portable computers with the keyboard behind a palm rest, and with a built-in pointing device (a trackball) in front of the keyboard.[60] The 1993 PowerBook 165c was Apple's first portable computer to feature a color screen, displaying 256 colors with 640 x 400 pixel resolution.[61] The second-generation of PowerBooks, the 68040-equipped 500 series, introduced the trackpad, integrated stereo speakers and built-in Ethernet to the laptop form factor in 1994.[62]

As for Mac OS, System 7 was a 32-bit rewrite from Pascal to C++ that introduced virtual memory, and improved the handling of color graphics, memory addressing, networking, and co-operative multitasking. Also during this time, the Macintosh began to shed the "Snow White" design language, along with the expensive consulting fees they were paying to Frogdesign, in favor of bringing the work in-house by establishing the Apple Industrial Design Group. They became responsible for to crafting a new look to go with the new operating system and all other Apple products.[63]

Despite these technical and commercial successes, Microsoft and Intel began to rapidly lower Apple's market share with the Windows 95 operating system and Pentium processors respectively. These significantly enhanced the multimedia capability and performance of IBM PC compatible computers, and brought Windows still closer to the

Mac GUI. Furthermore, Apple had created too many similar models that confused potential buyers. At one point Apple's product lineup was subdivided into Classic, LC, II, Quadra, Performa, and Centris models, with essentially the same computer being sold under a number of different names.[64] These models competed against the Macintosh clones, hardware manufactured by third-parties that ran Apple's System 7. This succeeded in increasing the Macintosh's market share somewhat and provided cheaper hardware for consumers, but hurt Apple financially as existing Apple customers began to buy cheaper clones while Apple shouldered the burden of developing the platform.

When Steve Jobs returned to Apple in 1997, he ordered that the OS that had been previewed as version 7.7 be branded Mac OS 8 (in place of the never-to-appear Copland OS). Since Apple had licensed only System 7 to third-parties, this move effectively ended the clone line. The decision caused significant financial losses for companies like Motorola, who produced the StarMax, Umax, who produced the SuperMac,[65] and Power Computing Corporation, who offered several lines Mac clones, including PowerWave, PowerTower, and PowerTower Pro.[66] These companies had invested substantial resources in creating their own Mac-compatible hardware.[67] Apple bought out Power Computing's license, but allowed Umax to continue selling Mac clones until their license expired, as they had a sizeable presence in the lower-end segment that Apple did not.

1998 to 2005: New beginnings

In 1998, a year after Steve Jobs had returned to the company, Apple introduced an all-in-one Macintosh called the iMac. Its translucent plastic case, originally Bondi blue and later many other colors, is considered an industrial design landmark of the late 1990s. The iMac did away with most of Apple's standard (and usually proprietary) connections, such as SCSI and ADB, in favor of two USB ports. It also had no internal floppy disk drive and instead included a CD-ROM drive for installing software,[3] [69] but was incapable of writing to CDs or other media without external third-party hardware. The iMac proved to be phenomenally successful, with 800,000 units sold in 139 days,[70] making the company an annual profit of US$309 million—Apple's first profitable year since Michael Spindler took over as CEO in 1995.[71] The iMacs's "blue and white" aesthetic was applied to the Power Macintosh, and then to a new product, the iBook. Introduced in July 1999, the iBook was Apple's first consumer-level laptop computer, filling the missing quadrant of Apple's "four-square product matrix" (*desktop* and *portable* products for both *consumers* and *professionals*).[72] More than 140,000 pre-orders were placed before it started shipping in September,[73] and by October it was as much a sales hit as the iMac.[74]

The original "Bondi Blue" iMac G3. Introduced in 1998, it led Apple's return to profitability. However, the associated mouse proved to be one of consumers' least favorite Apple products.[68]

In early 2001, Apple began shipping computers with CDRW drives for the first time.[75] Apple had been emphasizing the Mac's ability to play DVDs by including DVD-ROM and DVD-RAM drives as standard. Steve Jobs admitted that Apple had been "late to the party" on writable CD technology but felt that Macs could become a "digital hub" that linked and enabled an "emerging digital lifestyle".[76] Apple would later introduce an update to its iTunes music player software that could burn CDs, along with a controversial "Rip, Mix, Burn" advertising campaign that some[77] felt encouraged media piracy.[78] This accompanied the release of the iPod, Apple's first successful handheld device.

Apple continued to add new products to their lineup, such as the Power Mac G4 Cube,[79] the eMac for the education market and PowerBook G4 laptop for professionals. The original iMac used a G3 processor, but the G4 and then G5 chips were accompanied by successive new designs, dropping the array of colors in favor of white plastic. As of

2007 iMacs used aluminum enclosures. On January 11, 2005, Apple announced the release of the Mac Mini priced at US$499,[80] the least expensive Mac at its release.[81]

Mac OS continued to evolve up to version 9.2.2, including retrofits such as the addition of a nanokernel and support for Multiprocessing Services 2.0 in Mac OS 8.6.[82] Ultimately its dated architecture made replacement necessary. Initially developed in the Pascal programming language, it was substantially rewritten in C++ for System 7. From its beginnings on a 128k 8 MHz machine, it had grown to support Apple's latest 1 GHz G4-equipped Macs. But since its architecture was laid down, OS features like preemptive multitasking and protected memory had become feasible on the kind of hardware Apple manufactured – features that were already common on Apple's competition. As such, Apple introduced Mac OS X, a fully overhauled Unix-based successor to Mac OS 9, using Darwin, XNU, and Mach as foundations, and based on NEXTSTEP. Mac OS X was not released to the public until September 2000, as the Mac OS X Public Beta, with a revamped user interface Apple called "Aqua". At US$29.99, it allowed adventurous Mac users to sample Apple's new operating system and provide feedback for the actual release.[83] The initial release of Mac OS X, 10.0 (nicknamed Cheetah), was released on March 24, 2001. Older Mac OS applications could still run under early Mac OS X versions, using an environment called Classic. Subsequent releases of Mac OS X were 10.1 "Puma" (September 25, 2001), 10.2 "Jaguar" (August 24, 2002), 10.3 "Panther" (October 24, 2003) and 10.4 "Tiger" (April 29, 2005).

2006 onward: Intel era

Apple discontinued the use of PowerPC microprocessors in 2006. At WWDC 2005, Steve Jobs revealed this transition and also noted that Mac OS X was in development to run both on Intel and PowerPC architecture from the very beginning.[85] All new Macs now use x86 processors made by Intel, and some Macs were given new names to signify the switch.[86] Intel-based Macs can run pre-existing software developed for PowerPC using an emulator called Rosetta,[87] although at noticeably slower speeds than native programs, and the Classic environment is unavailable. Intel chips introduced the potential to run the Microsoft Windows operating system natively on Apple hardware, without emulation software such as Virtual PC. In March 2006, a group of hackers announced that they were able to run Windows XP on an Intel-based Mac. The group released their software as open source and has posted it for download on their website.[88] On April 5, 2006, Apple announced the public beta availability of their own Boot Camp software that allows owners of Intel-based Macs to install Windows XP on their machines; later versions added support for Windows Vista and Windows 7. Boot Camp became a standard feature in Mac OS X 10.5, while support for Classic was dropped from PowerPC Macs.[89] [90]

The MacBook Pro is the first Mac notebook to use an Intel processor. It was released at Macworld 2006.[84]

Starting in 2006, Apple's industrial design has shifted to favor aluminum, which was used in the construction of the first MacBook Pro. Glass was added in 2008 with the introduction of the unibody MacBook Pro. These materials are billed as environmentally friendly.[91] The iMac, MacBook Pro and MacBook Air lines use aluminum enclosures, and are now made of a single unibody.[92] [93] Chief designer Jonathan Ive continues to guide products towards a minimalist and simple feel,[94] [95] including the elimination of replaceable batteries in notebooks.[96] Multi-touch gestures from the iPhone's interface have been applied to the Mac line in the form of touch pads on notebooks and the Magic Mouse and Magic Trackpad for desktops.

In recent years, Apple has seen a significant boost in sales of Macs. This has been attributed, in part, to the success of the iPod and the iPhone, a halo effect whereby satisfied iPod or iPhone owners purchase more Apple equipment,[97] as well as the use of Intel microprocessors.[98] From 2001 to 2008, Mac sales increased continuously on an annual basis. Apple reported worldwide sales of 3.36 million Macs during the 2009 holiday season.[99] As of Mid-2011, Macintosh continues to enjoy rapid market share increase in the US, growing from 7.3% of all computer shipments in 2010 to 9.3% in 2011.[100]

On February 24, 2011, Apple was the first company to bring to market a computer that utilized Intel's new Thunderbolt (codename Light Peak) I/O interface. Using the same physical interface as a Mini DisplayPort, and backwards compatible with that standard, Thunderbolt boasts two-way transfer speeds of 10 Gbit/s.[101]

Product line

	Compact	Consumer	Professional
Desktop	**Mac mini** 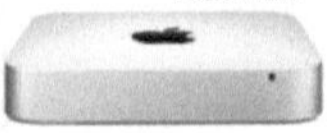 Entry-level; ships without keyboard, mouse, or monitor; uses Intel Core i5, or Intel Core i7 processors	**iMac**  All-in-one; available in 21.5" and 27" screen sizes; uses Intel Core i3, Intel Core i5, or Intel Core i7 processors	**Mac Pro**  Workstation desktop; highly customizable; Available with 4, 8 or 12 cores. Uses up-to two Intel Xeon X5650 "Westmere" (hexa-core) or Xeon X3460 "Nehalem" quad-core processors
Portable (MacBook)	**MacBook Air 11**  11.6" ultraportable with aluminum casing; uses Intel Core i5, or Intel Core i7 processors	**MacBook Air 13**  13.3" ultraportable with aluminum casing; uses Intel Core i5, or Intel Core i7 processors	**MacBook Pro**  13.3", 15.4" or 17" models with aluminum casing; uses Intel Core i5, or Intel Core i7 processors
Server	**Mac mini Server**  An additional Mac mini configuration that ships with Mac OS X Server installed.		**Mac Pro Server**  An additional Mac Pro server configuration that ships with Mac OS X Server installed.

Hardware and software

Hardware

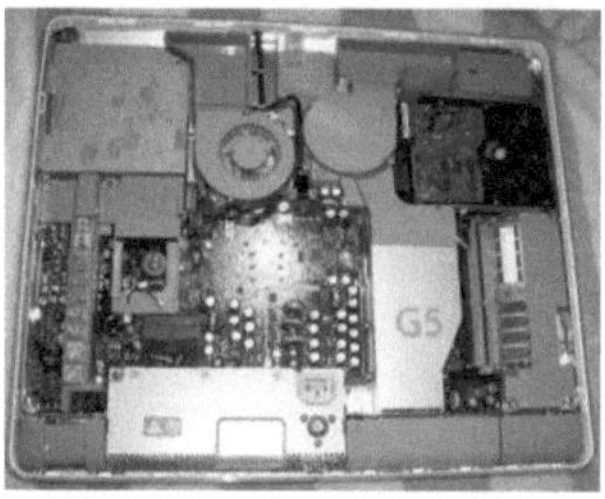
The internals of the original 20-inch iMac G5.

Apple directly sub-contracts hardware production to Asian original equipment manufacturers such as Asus, maintaining a high degree of control over the end product. By contrast, most other companies (including Microsoft) create software that can be run on hardware produced by a variety of third-parties, like Dell, HP/Compaq, and Lenovo. Consequently, the Macintosh buyer has comparably fewer options.

The current Mac product family uses Intel x86-64 processors. Apple introduced an emulator during the transition from PowerPC chips (called Rosetta), much as it did during the transition from Motorola 68000 architecture a decade earlier. The Macintosh is the only mainstream computer platform to have successfully transitioned to a new CPU architecture,[102] and has done so twice. All current Mac models ship with at least 2 GB RAM as standard. Current Mac computers use ATI Radeon or nVidia GeForce graphics cards as well as Intel graphics built into the main CPU. All current Macs (except for the MacBook Air and Mac Mini) ship with an optical media drive that includes a dual-function DVD/CD burner, called the SuperDrive. Macs include two standard data transfer ports: USB and FireWire (except for the MacBook Air and MacBook that do not include FireWire). MacBook Pro, iMac, and MacBook Air computers now also feature the "Thunderbolt" port, which Apple says can transfer data at speeds up to 10 gigabits per second. USB was introduced in the 1998 iMac G3 and is ubiquitous today,[3] while FireWire is mainly reserved for high-performance devices such as hard drives or video cameras. Starting with a new iMac G5 released in October 2005, Apple started to include built-in iSight cameras to appropriate models, and a media center interface called Front Row that can be operated by an Apple Remote or keyboard for accessing media stored on the computer.[103]

Apple was initially reluctant to embrace mice with multiple buttons and scroll wheels. Macs did not natively support multiple buttons, even from third parties, until Mac OS X arrived in 2001.[104] Apple continued to offer only single button mice, with wired and Bluetooth wireless versions, until August 2005, when it introduced the Mighty Mouse. While it looked like a traditional one-button mouse, it actually had four buttons and a scroll ball, capable of independent *x*- and *y*-axis movement.[105] A Bluetooth version followed in July 2006.[106] In October 2009, Apple introduced the Magic Mouse which uses multi-touch gesture recognition similar to the iPhone instead of a physical scroll wheel or ball.[107] It is available only in Bluetooth, and the Mighty Mouse (re-branded as "Apple Mouse") is available with a cord. Apple also features the "Magic Trackpad" as a means to control Macintosh desktop computers such as the iMac or Mac Pro. This was introduced in 2010.

Software

The original Macintosh was the first successful personal computer to use a graphical user interface devoid of a command line. It used a desktop metaphor, depicting real-world objects like documents and a trashcan as icons onscreen. The *System* software introduced in 1984 with the first Macintosh and renamed *Mac OS* in 1997, continued to evolve until version 9.2.2. In 2001, Apple introduced Mac OS X, based on Darwin and NEXTSTEP; its new features included the Dock and the Aqua user interface. During the transition, Apple included an emulator known as Classic allowing users to run Mac OS 9 applications under Mac OS X, version 10.4 and earlier on PowerPC machines. The most recent version is Mac OS X v10.7 "Lion." In addition to Lion, all new Macs are bundled with assorted Apple-produced applications, including iLife, the Safari web browser and the iTunes media player. Apple released Mac OS X 10.7 in 2010, which was made available in the summer of 2011. This operating system includes many new features such as: Mission Control, the Mac App Store (available to Mac OS X v10.6.6 "Snow Leopard."

users by software update),Launchpad, an application viewer and launcher akin to the iOS Home Screen, and Resume, a feature similar to the hibernate function found in Microsoft Windows.

Historically Mac OS X enjoyed a near-absence of the types of malware and spyware that affect Microsoft Windows users.[108] [109] [110] Mac OS X has a smaller usage share compared to Microsoft Windows (roughly 5% and 92%, respectively),[111] but it also has secure UNIX roots. Worms as well as potential vulnerabilities were noted in February 2006, that led some industry analysts and anti-virus companies to issue warnings that Apple's Mac OS X is not immune to malware.[112] Increasing market share coincided with additional reports of a variety of attacks.[113] Apple releases security updates for its software.[114] In early 2011 Mac OS X experienced a large increase in malware attacks,[115] and malware such as Mac Defender, MacProtector, MacGuard were seen as an increasing problem for Mac users. At first, the malware installer required the user to enter the administrative password but later versions were able to install without user input[116] Initially Apple support staff were instructed not to assist in the removal of the malware or admit the existence of the malware issue, but as the malware spread a support document was issued and Apple announced a OS X update to fix the problem. An estimated 100,000 users were affected.[117] [118]

Originally, the hardware architecture was so closely tied to the Mac OS operating system that it was impossible to boot an alternative operating system. The most common workaround, used even by Apple for A/UX, was to boot into Mac OS and then to hand over control to a program that took over the system and acted as a boot loader. This technique was no longer necessary with the introduction of Open Firmware-based PCI Macs, though it was formerly used for convenience on many Old World ROM systems due to bugs in the firmware implementation. Now, Mac hardware boots directly from Open Firmware (most PowerPC-based Macs) or EFI (all Intel-based Macs), and Macs are no longer limited to running just Mac OS X.

Following the release of the Intel-based Mac, third-party platform virtualization software such as Parallels Desktop, VMware Fusion, and VirtualBox began to emerge. These programs allow users to run Microsoft Windows or previously Windows-only software on Macs at near native speed. Apple also released Boot Camp and Mac-specific Windows drivers that help users to install Windows XP or Vista and natively dual boot between Mac OS X and Windows. Though not condoned by Apple, it is possible to run the Linux operating system using Boot camp or other virtualization workarounds.[119] [120]

Because Mac OS X is a UNIX-like operating system, borrowing heavily from FreeBSD, many applications written for Linux or BSD run on Mac OS X, often using X11. Apple's smaller market share than Microsoft's means that a smaller range of shareware is available, but many popular commercial software applications from large developers such as Microsoft Office and Adobe Photoshop are ported to both Mac OS and Windows. And much of open source software like the Firefox web browser and the OpenOffice.org office suite are cross-platform and run natively. But Mac OS X does not have as many games and "Just for fun" apps as Windows does, so some still prefer Windows.

Advertising

Apple hyped the introduction of the original Mac with their 1984 commercial that aired during the Super Bowl.[121] It was supplemented by a number of printed pamphlets and other TV ads demonstrating the new interface and emphasizing the mouse. Many more brochures for new models like the Macintosh Plus and the Performa followed. In the 1990s, Apple started the "What's on your PowerBook?" campaign, with print ads and television commercials featuring celebrities describing how the PowerBook helps them in their businesses and everyday lives. In 1995, Apple responded to the introduction of Windows 95 with several print ads and a television commercial demonstrating its disadvantages and lack of innovation. In 1997 the Think Different campaign introduced Apple's new slogan, and in 2002 the Switch campaign followed. The most recent advertising strategy by Apple is the Get a Mac campaign, with North American, UK and Japanese variants.[122] [123]

Today, Apple introduces new products at "special events" at the Apple Town Hall auditorium, and keynotes at the Apple Worldwide Developers Conference, and (formerly) trade shows like the Apple Expo and the MacWorld Expo.

The events typically draw a large gathering of media representatives and spectators, and are preceded by speculation about possible new products. In the past, special events have been used to unveil its desktop and notebook computers such as the iMac and MacBook, and other consumer electronic devices like the iPod, Apple TV, and iPhone, as well as provide updates on sales and market share statistics. Apple has begun to focus its advertising on its retail stores instead of these trade shows; the last MacWorld keynote was in 2009.[124]

Market share and user demographics

Since the introduction of the Macintosh, Apple has struggled to gain a significant share of the personal computer market. At first, the Macintosh 128K suffered from a dearth of available software compared to IBM's PC, resulting in disappointing sales in 1984 and 1985. It took 74 days for 50,000 units to sell.[125] Market share is measured by browser hits, sales and installed base. If using the browser metric, Mac market share has increased substantially in 2007.[126] If measuring market share by installed base, there were more than 20 million Mac users by 1997, compared to an installed base of around 340 million Windows PCs.[127] [128] Statistics from late 2003 indicate that Apple had 2.06 percent of the desktop share in the United States that had increased to 2.88 percent by Q4 2004.[129] As of October 2006, research firms IDC and Gartner reported that Apple's market share in the U.S. had increased to about 6 percent.[130] Figures from December 2006, showing a market share around 6 percent (IDC) and 6.1 percent (Gartner) are based on a more than 30 percent increase in unit sale from 2005 to 2006. The installed base of Mac computers is hard to determine, with numbers ranging from 5% (estimated in 2009)[131] to 16% (estimated in 2005).[132] Mac OS X's share of the OS market increased from 7.31% in December 2007 to 9.63% in December 2008, which is a 32% increase in market share during 2008, compared with a 22% increase during 2007.

By March 2011, OS X market share in North America had increased to slightly over 14%.[133] Whether the size of the Mac's market share and installed base is relevant, and to whom, is a hotly debated issue. Industry pundits have often called attention to the Mac's relatively small market share to predict Apple's impending doom, particularly in the early and mid 1990s when the company's future seemed bleakest. Others argue that market share is the wrong way to judge the Mac's success. Apple has positioned the Mac as a higher-end personal computer, and so it may be misleading to compare it to a budget PC.[134] Because the overall market for personal computers has grown rapidly, the Mac's increasing sales numbers are effectively swamped by the industry's expanding sales volume as a whole. Apple's small market share, then, gives the impression that fewer people are using Macs than did ten years ago, when exactly the opposite is true.[135] Soaring sales of the iPhone and iPad mean that the portion of Apple's profits represented by the Macintosh has declined in 2010, dropping to 24% from 46% two years earlier.[136] Others try to de-emphasize market share, citing that it is rarely brought up in other industries.[137] Regardless of the Mac's market share, Apple has remained profitable since Steve Jobs' return and the company's subsequent reorganization.[138] Notably, a report published in the first quarter of 2008 found that Apple had a 14% market share in the personal computer market in the US, including 66% of all computers over $1,000.[139] Market research indicates that Apple draws its customer base from a higher-income demographic than the mainstream personal computer market.[140]

See also

- Apple Inc. litigation
- Apple community
- History of computing hardware (1960s-present)
- List of Macintosh models by case type
- List of Macintosh models grouped by CPU type
- List of Macintosh software
- List of Macintosh software published by Microsoft
- Macintosh User Groups
- Mac gaming

- Reality distortion field
- Lilith (computer)

Notes

[1] "Define Macintosh" (http://dictionary.reference.com/browse/macintosh). Dictionary.com. . Retrieved April 11, 2010.

[2] Polsson, Ken (July 29, 2009). "Chronology of Apple Computer Personal Computers" (http://www.islandnet.com/~kpolsson/applehis/appl1984.htm). . Retrieved August 27, 2009. See May 3, 1984.

[3] Edwards, Benj (August 15, 2008). "Eight ways the iMac changed computing" (http://www.macworld.com/article/135017/2008/08/imacanniversary.html). Macworld. . Retrieved August 27, 2009.

[4] "Xserve Transition Guide" (http://images.apple.com/xserve/pdf/L422277A_Xserve_Guide.pdf). November 5, 2010. . Retrieved November 5, 2010.

[5] Raskin, Jef (1996). "Recollections of the Macintosh project" (http://mxmora.best.vwh.net/JefRaskin.html). *Articles from Jef Raskin about the history of the Macintosh.*. . Retrieved November 27, 2008.

[6] Apple confidential 2.0: the definitive history of the world's most colorful company, Owen W. Linzmayer, ISBN 978-1-59327-010-0

[7] Hertzfeld, Andy. "The father of the Macintosh" (http://www.folklore.org/StoryView.py?project=Macintosh&story=The_Father_of_The_Macintosh.txt&showcomments=1#comments). Folklore.org. . Retrieved April 24, 2006.

[8] Crow, George. "The Original Macintosh" (http://www.folklore.org/ProjectView.py?project=Macintosh&characters=George Crow). Folklore.org. . Retrieved April 28, 2010.

[9] Kottke, Dan. "The Original Macintosh" (http://www.folklore.org/ProjectView.py?project=Macintosh&characters=Dan Kottke). Folklore.org. . Retrieved April 28, 2010.

[10] Manock, Jerry. "The Original Macintosh" (http://web.archive.org/web/20070929111006/http://folklore.org/ProjectView.py?project=Macintosh&characters=Jerry Manock). Folklore.org. Archived from the original (http://folklore.org/ProjectView.py?project=Macintosh&characters=Jerry Manock) on September 29, 2007. . Retrieved April 28, 2010.

[11] Kawasaki, Guy (January 26, 2009). "Macintosh 25th Anniversary Reunion: Where Did Time Go?" (http://blog.guykawasaki.com/2009/01/twenty-five-yea.html#axzz0mQb8tIS0). . Retrieved April 28, 2010.

[12] Hertzfeld, Andy. "Five different Macintoshes" (http://www.folklore.org/StoryView.py?story=Five_Different_Macs.txt). Folklore.org. . Retrieved April 24, 2006.

[13] Horn, Bruce. "On Xerox, Apple and Progress" (http://www.folklore.org/StoryView.py?project=Macintosh&story=On_Xerox,_Apple_and_Progress.txt). Folklore.org. . Retrieved February 3, 2007.

[14] Tracy, Ed. "History of computer design: Snow White" (http://www.landsnail.com/apple/local/design/design2.html). Landsnail.com. . Retrieved April 24, 2006.

[15] Hertzfeld, Andy. "The End Of An Era" (http://www.folklore.org/StoryView.py?project=Macintosh&story=The_End_Of_An_Era.txt). *folklore.org*. .

[16] Spector, G (September 24, 1985). "Apple's Jobs Starts New Firm, Targets Education Market". *PCWeek*: p. 109.

[17] "Apple Computer, Inc. Finalizes Acquisition of NeXT Software Inc." (http://web.archive.org/web/19990117075346/http://product.info.apple.com/pr/press.releases/1997/q2/970207.pr.rel.next.html). Apple. February 7, 1997. Archived from the original (http://product.info.apple.com/pr/press.releases/1997/q2/970207.pr.rel.next.html) on January 17, 1999. . Retrieved April 27, 2010.

[18] Dennis, Rob (2011-12-30). "Fremont mayor Bob Wasserman dead at 77" (http://www.insidebayarea.com/news/ci_19641994). *The Argus (Fremont)*. . Retrieved 2012-01-21.

[19] "Apple Macintosh 18 Page Brochure" (http://www.digibarn.com/collections/ads/apple-mac/index.htm). DigiBarn Computer Museum. . Retrieved April 24, 2006.

[20] Linzmayer, Owen W. (2004). *Apple Confidential 2.0* (http://www.oweninnk.com). No Starch Press. p. 113. ISBN 1-59327-010-0. .

[21] Maney, Kevin (January 28, 2004). "Apple's '1984' Super Bowl commercial still stands as watershed event" (http://www.usatoday.com/tech/columnist/kevinmaney/2004-01-28-maney_x.htm). *USA Today*. . Retrieved April 11, 2010.

[22] Leopold, Todd (February 3, 2006). "Why 2006 isn't like '1984'" (http://edition.cnn.com/2006/SHOWBIZ/02/02/eye.ent.commercials/). CNN. . Retrieved May 10, 2008.

[23] Cellini, Adelia (January 2004). "The Story Behind Apple's '1984' TV commercial: Big Brother at 20" (http://web.archive.org/web/20080626112220/http://findarticles.com/p/articles/mi_hb197/is_200401/ai_n5556112). *MacWorld 21.1, page 18*. Archived from the original (http://findarticles.com/p/articles/mi_hb197/is_200401/ai_n5556112) on June 26, 2008. . Retrieved May 9, 2008.

[24] Long, Tony (January 22, 2007). "Jan. 22, 1984: Dawn of the Mac" (http://www.wired.com/science/discoveries/news/2007/01/72496). Wired. . Retrieved April 11, 2010.

[25] Kahney, Leander (January 6, 2004). "We're All Mac Users Now" (http://www.wired.com/gadgets/mac/news/2004/01/61730). Wired. . Retrieved April 11, 2010.

[26] Polsson, Ken. "Chronology of Apple Computer Personal Computers" (http://www.islandnet.com/~kpolsson/applehis/appl1984.htm). . Retrieved November 18, 2007.

[27] Beamer, Scott (January 13, 1992). "For Lotus, third time's the charm" (http://www.accessmylibrary.com/article-1G1-11721498/lotus-third-time-charm.html). MacWEEK. . Retrieved June 23, 2010.

[28] Hornby, Thomas (October 2, 2006). "Apple's Worst Business Decisions" (http://www.osnews.com/story/16036/Apples-Worst-Business-Decisions/). OS News. . Retrieved December 24, 2007.
[29] "1984 *Newsweek* Macintosh ads" (http://www.guidebookgallery.org/ads/magazines/macos/macos10-newsweek). GUIdebook, Newsweek. . Retrieved April 24, 2006.
[30] "Inflation Calculator" (http://data.bls.gov/cgi-bin/cpicalc.pl). Bureau of Labor Statistics. . Retrieved May 14, 2010.
[31] Spring, Michael B. (1991). *Electronic printing and publishing: the document processing revolution* (http://books.google.com/?id=_MV46vFUrI4C&pg=PA125&dq=Desktop+publishing+history+of&cd=5#v=onepage&q=Desktop publishing history of). CRC Press. pp. 125–126. ISBN 978-0-8247-8544-4. .
[32] Technical specifications of Macintosh 512K from Apple's knowledge base (http://support.apple.com/kb/SP187) and from EveryMac.com (http://www.everymac.com/systems/apple/mac_classic/stats/mac_512k.html). Retrieved June 23, 2010.
[33] Technical specifications of Macintosh Plus from Apple's knowledge base (http://support.apple.com/kb/SP190) and from EveryMac.com (http://www.everymac.com/systems/apple/mac_classic/stats/mac_plus.html). Retrieved June 23, 2010.
[34] Technical specifications of Macintosh II from Apple's knowledge base (http://support.apple.com/kb/SP193) and from EveryMac.com (http://www.everymac.com/systems/apple/mac_ii/stats/mac_ii.html). Retrieved June 23, 2010.
[35] "Apple Macintosh II" (http://www.old-computers.com/museum/computer.asp?st=1&c=160). Old Computers On-line Museum. . Retrieved December 23, 2007.
[36] "Macintosh II Family: Fan Regulator Voids Warranty" (http://support.apple.com/kb/TA40560). Apple. July 2, 1992. . Retrieved December 23, 2007.
[37] Technical specifications of Macintosh SE from Apple's knowledge base (http://support.apple.com/kb/SP191) and from EveryMac.com (http://www.everymac.com/systems/apple/mac_classic/stats/mac_se.html). Retrieved June 23, 2010.
[38] "Apple Service Source: Apple SE/30" (http://www.apple-collection.com/CarPos/macse30.pdf). . Retrieved November 16, 2010.
[39] Hearm, Bob (2003). "A Brief History of ClarisWorks" (http://www-swiss.ai.mit.edu/~bob/clarisworks.php). MIT Project on Mathematics and Computation. . Retrieved December 24, 2007.
[40] Free Software Foundation (June 11, 1988). "Special Report: Apple's New Look and Feel" (http://www.gnu.org/bulletins/bull5.html#SEC9). *GNU's Bulletin* **1** (5). . Retrieved April 25, 2006.
[41] Free Software Foundation (1995-01). "End of Apple Boycott" (http://www.gnu.org/bulletins/bull18.html#SEC13). *GNU's Bulletin* **1** (18). . Retrieved April 25, 2006.
[42] Technical specifications of Macintosh IIx from Apple's knowledge base (http://support.apple.com/kb/SP194) and from EveryMac.com (http://www.everymac.com/systems/apple/mac_ii/stats/mac_iix.html). Retrieved June 23, 2010.
[43] Technical specifications of Macintosh IIcx from Apple's knowledge base (http://support.apple.com/kb/{{{appleNumber}}}) and from EveryMac.com (http://www.everymac.com/systems/apple/apple/mac_ii/stats/mac_iicx.html). Retrieved June 23, 2010.
[44] Technical specifications of Macintosh SE/30 from Apple's knowledge base (http://support.apple.com/kb/SP195) and from EveryMac.com (http://www.everymac.com/systems/apple/mac_classic/stats/mac_se30.html). Retrieved June 23, 2010.
[45] Technical specifications of Macintosh IIci from Apple's knowledge base (http://support.apple.com/kb/SP197) and from EveryMac.com (http://www.everymac.com/systems/apple/mac_ii/stats/mac_iici.html). Retrieved June 23, 2010.
[46] Knight, Dan (2001-01). "32-bit Addressing on Older Macs" (http://lowendmac.com/trouble/32bit.shtml). Low End Mac. . Retrieved December 24, 2007.
[47] Technical specifications of Macintosh Portable from Apple's knowledge base (http://support.apple.com/kb/SP140) and from EveryMac.com (http://www.everymac.com/systems/apple/powerbook/stats/mac_portable.html). Retrieved June 23, 2010.
[48] Technical specifications of Macintosh IIfx from Apple's knowledge base (http://support.apple.com/kb/SP203) and from EveryMac.com (http://www.everymac.com/systems/apple/mac_ii/stats/mac_iifx.html). Retrieved June 23, 2010.
[49] Fisher, Lawrence M. (October 15, 1990). "Less-Costly Apple Line To Be Presented Today" (http://query.nytimes.com/gst/fullpage.html?res=9C0CEFD9163CF936A25753C1A966958260). *The New York Times*. . Retrieved January 16, 2008.
[50] Technical specifications of Macintosh LC from Apple's knowledge base (http://support.apple.com/kb/SP205) and from EveryMac.com (http://www.everymac.com/systems/apple/mac_lc/stats/mac_lc.html). Retrieved June 24, 2010.
[51] Technical specifications of Macintosh IIsi from Apple's knowledge base (http://support.apple.com/kb/SP199) and from EveryMac.com (http://www.everymac.com/systems/apple/mac_ii/stats/mac_iisi.html). Retrieved June 24, 2010.
[52] Fisher, Lawrence M. (January 18, 1991). "I.B.M. Surprises Wall Street With Strong Quarterly Net; Apple Posts 20.6% Rise" (http://query.nytimes.com/gst/fullpage.html?res=9D0CE2DA163BF93BA25752C0A967958260). *The New York Times*. . Retrieved January 16, 2008.
[53] Technical specifications of Macintosh Classic II from Apple's knowledge base (http://support.apple.com/kb/SP204) and from EveryMac.com (http://www.everymac.com/systems/apple/mac_classic/stats/mac_classic_ii.html). Retrieved June 24, 2010.
[54] Technical specifications of Macintosh LC II from Apple's knowledge base (http://support.apple.com/kb/SP206) and from EveryMac.com (http://www.everymac.com/systems/apple/mac_lc/stats/mac_lc_ii.html). Retrieved June 24, 2010.
[55] Technical specifications of Macintosh Quadra 700 from Apple's knowledge base (http://support.apple.com/kb/SP235) and from EveryMac.com (http://www.everymac.com/systems/apple/mac_quadra/stats/mac_quadra_700.html). Retrieved June 24, 2010.
[56] Technical specifications of Macintosh Quadra 900 from Apple's knowledge base (http://support.apple.com/kb/mac_quadra/stats/mac_quadra_900.html) and from EveryMac.com (http://www.everymac.com/systems/apple/mac_quadra/stats/mac_quadra_900.html). Retrieved June 24, 2010.

[57] Hornby, Thomas (January 3, 2005). "Apple's Transition to PowerPC put in perspective" (http://www.kaomso.com/FullStory.php?TheStory=78). Kaomso. . Retrieved December 24, 2007.

[58] Polsson, Ken (December 16, 2007). "Chronology of Apple Computer Personal Computers" (http://www.islandnet.com/~kpolsson/applehis/appl1994.htm). . Retrieved December 24, 2007.

[59] Polsson, Ken. "Chronology of Apple Computer Personal Computers" (http://www.islandnet.com/~kpolsson/applehis/appl1990.htm). . Retrieved November 18, 2007.

[60] Jade, Kasper (February 16, 2007). "Apple to re-enter the sub-notebook market" (http://www.appleinsider.com/articles/07/02/16/apple_to_re_enter_the_sub_notebook_market.html). AppleInsider. . Retrieved December 24, 2007.

[61] Technical specifications of PowerBook 165c from Apple's knowledge base (http://support.apple.com/kb/SP148) and from EveryMac.com (http://www.everymac.com/systems/apple/powerbook/stats/mac_powerbook165c.html). Retrieved June 24, 2010.

[62] Technical specifications of PowerBook 520 from Apple's knowledge base (http://support.apple.com/kb/SP160) and from EveryMac.com (http://www.everymac.com/systems/apple/powerbook/stats/mac_powerbook520.html). Retrieved November 12, 2010.

[63] Kunkel, Paul (October 1, 1997). *AppleDesign: The work of the Apple Industrial Design Group*. Rick English (photographs). New York City: Graphis Inc.. ISBN 1-888001-25-9.

[64] Apple Computer (June 19, 1995). "Macintosh Centris, Quadra 660AV: Description (Discontinued)" (http://docs.info.apple.com/article.html?artnum=12707). . Retrieved December 24, 2007.

[65] EveryMac.com (October 27, 2009). "Umax Mac Clones (MacOS-Compatible Systems)" (http://www.everymac.com/systems/umax/index-umax-supermac-mac-clones.html). . Retrieved November 11, 2009.

[66] EveryMac.com (October 27, 2009). "PowerComputing Mac Clones (MacOS-Compatible Systems)" (http://www.everymac.com/systems/powercc/index-power-computing-mac-clones.html). . Retrieved November 11, 2009.

[67] Knight, Dan (August 30, 2007). "1997: Apple Squeezes Mac Clones Out of the Market" (http://lowendmac.com/musings/mm07/0830.html). Low End Mac. . Retrieved December 24, 2007.

[68] Engst, Adam (January 23, 2009). "The six worst Apple products of all time" (http://www.macworld.com/article/138404/2009/01/macat25_worstproducts.html). Macworld. . Retrieved May 14, 2010.

[69] Technical specifications of iMac G3 from Apple's knowledge base (http://support.apple.com/kb/SP136) and from EveryMac.com (http://www.everymac.com/systems/apple/imac/stats/imac_ab.html). Retrieved June 24, 2010.

[70] "800,000 iMacs Sold in First 139 Days" (http://www.apple.com/ca/press/1999/01/iMac_Sales.html). Apple. January 5, 1999. . Retrieved December 23, 2007.

[71] Markoff, John (October 15, 1998). "COMPANY REPORTS; Apple's First Annual Profit Since 1995" (http://query.nytimes.com/gst/fullpage.html?res=9D0CE6D8123AF936A25753C1A96E958260). *The New York Times*. . Retrieved December 23, 2007.

[72] "iBook: An iMac to Go" (http://db.tidbits.com/article/5487?print_version=1). .

[73] "Apple Averages Three Thousand iBooks Per Day In Pre-orders!" (http://www.macobserver.com/news/99/august/990831/140000ibooks.html). The Mac Observer. August 31, 1999. . Retrieved December 24, 2007.

[74] "PC Data Ranks iBook Number One Portable in U.S." (http://www.apple.com/pr/library/2000/jan/25ibook.html). Apple. January 25, 2000. . Retrieved December 18, 2007.

[75] "Speed, Song Highlight Apple Product Announcements" (http://pcworld.about.com/news/Jan092001id37951.htm). .

[76] "Apple picks up the beat with CD-RW drives" (http://news.cnet.com/Apple-picks-up-the-beat-with-CD-RW-drives/2100-1040_3-249578.html). .

[77] <http://www.macworld.com/article/3744/2002/03/eisner2.html>

[78] "The Complete iTunes History – SoundJam MP to iTunes 9" (http://www.maclife.com/article/feature/complete_itunes_history_soundjam_mp_itunes_9). .

[79] "About the Macintosh Cube" (http://manuals.info.apple.com/en/PowerMacG4_CubeAbout.PDF) (PDF). Apple. 2000. . Retrieved October 9, 2008.

[80] Markoff, John; Hansell, Saul (January 12, 2005). "Apple Changes Course With Low-Priced Mac" (http://www.nytimes.com/2005/01/12/technology/12apple.html). *The New York Times*. . Retrieved January 16, 2006.

[81] *Apple unveils low-cost 'Mac mini'* (http://news.bbc.co.uk/1/hi/technology/4162009.stm). BBC News. January 11, 2005. . Retrieved April 28, 2010

[82] "Apple Developer Connection – Overview of the PowerPC System Software" (http://developer.apple.com/documentation/mac/PPCSoftware/PPCSoftware-12.html#HEADING12-0). Apple. . Retrieved May 11, 2009.

[83] Biersdorfer, J.D. (September 14, 2000). "Apple Breaks The Mold" (http://query.nytimes.com/gst/fullpage.html?res=9F01E2DD1538F937A2575AC0A9669C8B63). *The New York Times*. . Retrieved January 16, 2008.

[84] Technical specifications of MacBook Pro from Apple's knowledge base (http://support.apple.com/kb/SP36) and from EveryMac.com (http://www.everymac.com/systems/apple/macbook_pro/stats/macbook_pro_1.83.html). Retrieved June 24, 2010.

[85] "Apple to Use Intel Microprocessors Beginning in 2006" (http://www.apple.com/pr/library/2005/jun/06intel.html). Apple. June 6, 2005. . Retrieved May 14, 2010.

[86] Michaels, Philip (January 2, 2010). "Apple's most significant products of the decade" (http://www.macworld.com/article/145380/2010/01/10significantapplemoves.html). Macworld. . Retrieved May 14, 2010.

[87] "WWDC 2005 Keynote Live Update" (http://www.macworld.com/article/45157/2005/06/liveupdate.html). Macworld. June 6, 2005. . Retrieved May 14, 2010.

[88] "Hackers get Windows XP to run on a Mac" (http://www.msnbc.msn.com/id/11885495/). MSNBC (AP). March 17, 2006. . Retrieved April 24, 2006.
[89] "Boot Camp" (http://www.apple.com/support/bootcamp/). Apple. . Retrieved May 17, 2010.
[90] Dalrymple, Jim (March 5, 2006). "New Apple software lets Intel Macs boot Windows" (http://www.macworld.com/article/50204/2006/04/bootcamp.html). Macworld. . Retrieved May 14, 2010.
[91] "The story behind Apple's environmental footprint." (http://www.apple.com/environment/). Apple. . Retrieved January 24, 2011.
[92] "New MacBook Family Redefines Notebook Design" (http://www.apple.com/pr/library/2008/10/14macbook.html). Apple. October 14, 2008. . Retrieved December 23, 2009.
[93] Lam, Brian (October 20, 2009). "Apple iMac Hands On" (http://gizmodo.com/5385841/apple-imac-hands-on). *Gizmodo*. . Retrieved August 18, 2010.
[94] Kahney, Leander (June 25, 2003). "Design According to Ive" (http://www.wired.com/culture/design/news/2003/06/59381?currentPage=all). Wired. . Retrieved December 23, 2009.
[95] Nosowitz, Dan (November 7, 2009). "Watch Jonathan Ive's Segment in *Objectified*" (http://gizmodo.com/5399420/watch-jonathan-ives-segment-in-objectified). Gizmodo. . Retrieved December 23, 2009.
[96] "Apple Updates MacBook Pro Family with New Models & Innovative Built-in Battery for Up to 40% Longer Battery Life" (http://www.apple.com/pr/library/2009/06/08mbp.html). Apple. June 8, 2009. . Retrieved December 23, 2009.
[97] Apple Aims for the Masses With a Cheaper iPhone (http://www.nytimes.com/2008/06/10/technology/10apple.html)
[98] Windows or Mac? Apple Says Both (http://query.nytimes.com/gst/fullpage.html?res=9506EED81030F935A35757C0A9609C8B63&pagewanted=all)
[99] "Apple Reports First Quarter Results" (http://www.apple.com/pr/library/2010/01/25results.html). Apple. January 25, 2009. .
[100] "Apple's Mac market share grows 18.9% in first quarter" (http://www.loopinsight.com/2011/04/13/apples-mac-market-share-grows-18-9-in-first-quarter/). Loopinsight.com. April 13, 2011. . Retrieved July 5, 2011.
[101] "Thunderbolt Technology: The Fastest Data Connection to Your PC Just Arrived" (http://newsroom.intel.com/community/intel_newsroom/blog/2011/02/24/thunderbolt-technology-the-fastest-data-connection-to-your-pc-just-arrived). .
[102] Dvorak, John C. (March 18, 2003). "Apple Switch" (http://www.pcmag.com/article2/0,2817,933453,00.asp). *PC Magazine*. .
[103] "Apple Introduces the New iMac G5" (http://www.apple.com/pr/library/2005/oct/12imac.html). Apple. October 12, 2005. . Retrieved July 12, 2006.
[104] "Eek! A Two-Button Mac Mouse?" (http://www.wired.com/science/discoveries/news/2000/10/39773). Wired. October 30, 2000. . Retrieved December 23, 2009.
[105] "Apple Introduces Mighty Mouse" (http://www.apple.com/pr/library/2005/aug/02mightymouse.html). Apple. August 2, 2005. . Retrieved July 12, 2006.
[106] "Apple Debuts Wireless Mighty Mouse" (http://www.apple.com/pr/library/2006/jul/25mightymouse.html). Apple. July 25, 2006. . Retrieved November 30, 2007.
[107] "Apple Introduces Magic Mouse—The World's First Multi-Touch Mouse" (http://www.apple.com/pr/library/2009/10/20magicmouse.html). Apple. October 20, 2009. . Retrieved December 24, 2009.
[108] Welch, John (January 6, 2007). "Review: Mac OS X Shines In Comparison With Windows Vista" (http://www.informationweek.com/news/showArticle.jhtml?articleID=196800670&pgno=4). Information Week. . Retrieved February 5, 2007.
[109] Granneman, Scott (October 6, 2003). "Linux vs. Windows Viruses" (http://www.theregister.co.uk/2003/10/06/linux_vs_windows_viruses/). The Register. . Retrieved February 5, 2007.
[110] Gruber, John (June 4, 2004). "Broken Windows" (http://daringfireball.net/2004/06/broken_windows). Daring Fireball. . Retrieved April 24, 2006.
[111] "Operating System Market Share" (http://marketshare.hitslink.com/operating-system-market-share.aspx?qprid=8). September 2009. . Retrieved April 10, 2009.
[112] Roberts, Paul (February 21, 2006). "New Safari Flaw, Worms Turn Spotlight on Apple Security" (http://www.eweek.com/article2/0,1895,1929342,00.asp). eWeek. . Retrieved November 23, 2007.
[113] Conneally, Tim (August 28, 2009). "'Macs don't get viruses' myth dissolves before public's eyes" (http://www.betanews.com/article/Macs-dont-get-viruses-myth-dissolves-before-publics-eyes/1251493625). BetaNews.com. .
[114] "Apple security updates" (http://support.apple.com/kb/HT1222). Apple. January 21, 2009. . Retrieved January 29, 2009.
[115] Grimes, Roger A. (May 23, 2011). "7 questions about the Mac malware scare | Security" (http://www.infoworld.com/d/security/7-questions-about-the-mac-malware-scare-811). InfoWorld. . Retrieved July 5, 2011.
[116] "Mac Security Boasts Threatened by Malware Surge – International Business Times" (http://www.ibtimes.com/articles/152612/20110526/mac-security-boasts-threatened-by-malware-surge.htm). Ibtimes.com. May 26, 2011. . Retrieved July 5, 2011.
[117] Trenholm, Rich (May 20, 2011). "Apple tells support staff not to confirm Mac Defender infections" (http://crave.cnet.co.uk/software/apple-tells-support-staff-not-to-confirm-mac-defender-infections-50003832/). UK: Crave.. . Retrieved July 5, 2011.
[118] Seltzer, Larry (May 25, 2011). "Mac Defender 2.0 Released – Security Watch" (http://blogs.pcmag.com/securitywatch/2011/05/mac_defender_20_released.php). *PC Mag*. . Retrieved July 5, 2011.
[119] Lucas, Paul (June 4, 2005). "Paul J. Lucas's Mac Mini running Linux" (http://homepage.mac.com/pauljlucas/personal/macmini/index.html). . Retrieved December 23, 2009.

[120] Hoover, Lisa (April 11, 2008). "Virtualization Makes Running Linux a Snap" (http://ostatic.com/blog/virtualization-makes-running-linux-a-snap). . Retrieved December 23, 2009.
[121] Pogue, David; Joseph Schorr (1993). *Macworld Macintosh SECRETS*. San Mateo: IDG Books Worldwide, Inc.. p. 251. ISBN 1-56884-025-X.
[122] "Get a Mac advertisements" (http://www.apple.com/uk/getamac/). Apple. . Retrieved January 29, 2007.
[123] "Get a Mac" (http://www.apple.com/jp/getamac/). Apple. . Retrieved February 3, 2007.
[124] "Apple Announces Its Last Year at Macworld" (http://www.apple.com/pr/library/2008/12/16macworld.html). Apple. December 16, 2008. . Retrieved March 30, 2009.
[125] Polsson, Ken (July 29, 2009). "Chronology of Apple Computer Personal Computers" (http://www.islandnet.com/~kpolsson/applehis/appl1984.htm). . Retrieved August 27, 2009. See April 7, 1984.
[126] "Trends in Mac market share" (http://arstechnica.com/journals/apple.ars/2007/04/05/trends-in-mac-market-share). Ars Technica. April 5, 2009. . Retrieved August 27, 2009.
[127] "Apple Developer News, No. 87" (http://developer.apple.com/adcnews/pastissues/devnews121997.html#stats). Apple Computer. December 19, 1997. . Retrieved April 24, 2006.
[128] "Nearly 600 Million Computers-in-Use in Year 2000" (http://www.c-i-a.com/pr1198.htm). Computer Industry Almanac Inc. November 3, 1998. . Retrieved June 1, 2006.
[129] Dalrymple, Jim (April 20, 2005). "Apple desktop market share on the rise; will the Mac mini, iPod help?" (http://www.macworld.com/news/2005/03/20/marketshare/index.php). Macworld. . Retrieved April 24, 2006.
[130] Dalrymple, Jim (October 19, 2006). "Apple's Mac market share tops 5% with over 30% growth" (http://www.macworld.com/news/2006/10/19/marketshare/index.php). Macworld. . Retrieved December 22, 2006.
[131] "Operating System Market Share" (http://marketshare.hitslink.com/operating-system-market-share.aspx?qprid=8). Hitslink. July 2009. . Retrieved August 27, 2009.
[132] MacDailyNews (June 15, 2005). "16% of computer users are unaffected by viruses, malware because they use Apple Macs" (http://macdailynews.com/index.php/weblog/comments/5933/). . Retrieved April 24, 2006.
[133] (http://9to5mac.com/2011/03/17/top-10-mac-countries-by-market-share-united-states-is-3/)
[134] Gruber, John (July 23, 2003). "Market Share" (http://daringfireball.net/2003/07/market_share). Daring Fireball. . Retrieved April 24, 2006.
[135] Brockmeier, Joe (May 13, 2003). "What Will It Take To Put Apple Back on Top?" (http://www.newsfactor.com/perl/story/21499.html). NewsFactor Magazine online. . Retrieved April 24, 2006.
[136] "Despite growing sales, Mac's share of Apple profits wanes" (http://www.thenewstribune.com/2010/10/21/1390638/despite-growing-sales-macs-share.html). .
[137] Toporek, Chuck (August 22, 2001). "Apple, Market Share, and Who Cares?" (http://www.oreillynet.com/mac/blog/2001/08/apple_market_share_and_who_car.html). O'Reilly macdevcenter.com. . Retrieved April 24, 2006.
[138] Spero, Ricky (July 14, 2004). "Apple Posts Profit of $61 million; Revenue Jumps 30%" (http://www.macobserver.com/stockwatch/2004/07/14.1.shtml). The Mac Observer. . Retrieved April 24, 2006.
[139] Wilcox, Joe. "Macs Defy Windows' Gravity" (http://blogs.eweek.com/applewatch/content/channel/macs_defy_windows-gravity.html). Apple Watch. . Retrieved May 19, 2008.
[140] Fried, Ian (July 12, 2002). "Are Mac users smarter?" (http://news.com.com/2100-1040-943519.html). news.com. . Retrieved April 24, 2006.

References

- Apple & Raskin, Jef (1992). *Macintosh Human Interface Guidelines*. Addison-Wesley Professional. ISBN 0-201-62216-5.
- Apple. "Press release Library" (http://www.apple.com/pr/library/). Retrieved November 18, 2007.
- Deutschman, Alan (2001). *The Second Coming of Steve Jobs*. Broadway. ISBN 0-7679-0433-8.
- Hertzfeld, Andy. "folklore.org: Macintosh stories" (http://folklore.org/index.py). Retrieved April 24, 2006.
- Hertzfeld, Andy (2004). *Revolution in the Valley*. O'Reilly Books. ISBN 0-596-00719-1.
- Kahney, Leander (2004). *The Cult of Mac*. No Starch Press. ISBN 1-886411-83-2.
- Kawasaki, Guy (1989). *The Macintosh Way*. Scott Foresman Trade. ISBN 0-673-46175-0.
- Kelby, Scott (2002). *Macintosh... The Naked Truth*. New Riders Press. ISBN 0-7357-1284-0.
- Knight, Dan (2005). "Macintosh History: 1984" (http://lowendmac.com/history/1984dk.shtml). Retrieved April 24, 2006.
- Levy, Steven (2000). *Insanely Great: The Life and Times of Macintosh, the Computer That Changed Everything*. Penguin Books. ISBN 0-14-029177-6.
- Linzmayer, Owen (2004). *Apple Confidential 2.0*. No Starch Press. ISBN 1-59327-010-0.

- Page, Ian (2007). "MacTracker Macintosh model database 4.3.1" (http://www.mactracker.ca/). Retrieved November 31, 2007.
- Sanford, Glen (2006). "Apple History" (http://www.apple-history.com/). Retrieved April 24, 2006.
- Singh, Amit (2005). "A History of Apple's Operating Systems" (http://www.kernelthread.com/mac/oshistory/). Retrieved April 24, 2006.

External links

- Official website (http://http://www.apple.com/mac/)
- Making the Macintosh: Technology and Culture in Silicon Valley (http://library.stanford.edu/mac/index.html)

Microsoft_Windows

Windows	
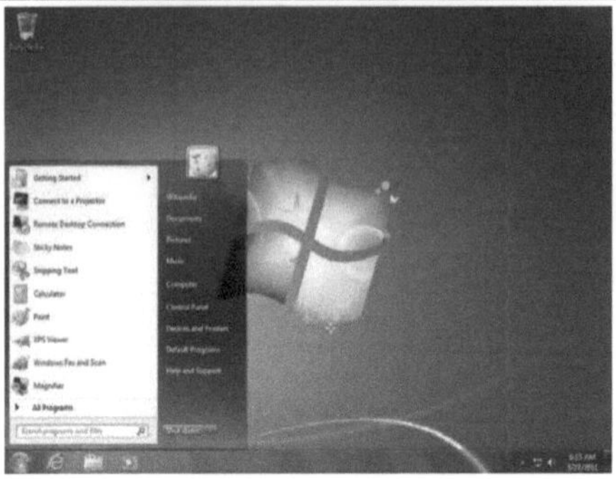 The latest Windows release, Windows 7, showing desktop and start menu	
Company / developer	Microsoft
Programmed in	C, C++ and Assembly language[1]
OS family	Windows 9x, Windows CE and Windows NT
Working state	Publicly released
Source model	Closed source / Shared source
Initial release	November 20, 1985 (as Windows 1.0)
Latest stable release	[[Windows 7 [2]], Windows Server 2008 R2 NT 6.1 (Build 7601: Service Pack 1)] (February 22, 2011) [3] [±]
Latest unstable release	[[Windows 8 [4]] NT 6.2 (Build 8250)] [5] [±]
Marketing target	Personal computing
Available language(s)	Multilingual (listing of available Windows 7 language packs [6])
Update method	Windows Update, Windows Anytime Upgrade
Supported platforms	ARM, IA-32, x86-64 and Itanium
Kernel type	Hybrid (Windows NT family), DOS (16-bit Windows and Windows 9x/ME series)
Default user interface	Graphical (Windows Shell)
License	Proprietary commercial software
Official website	[windows.microsoft.com windows.microsoft.com]

Microsoft Windows is a series of graphical interface operating systems developed, marketed, and sold by Microsoft.

Microsoft introduced an operating environment named *Windows* on November 20, 1985 as an add-on to MS-DOS in response to the growing interest in graphical user interfaces (GUIs).[7] Microsoft Windows came to dominate the world's personal computer market, overtaking Mac OS, which had been introduced in 1984.

The most recent client version of Windows is Windows 7; the most recent server version is Windows Server 2008 R2; the most recent mobile version is Windows Phone 7.5.

Versions

The term *Windows* collectively describes any or all of several generations of Microsoft operating system products. These products are generally categorized as follows:

The classic Windows logo, used until the release of Windows XP in 2001

Early versions

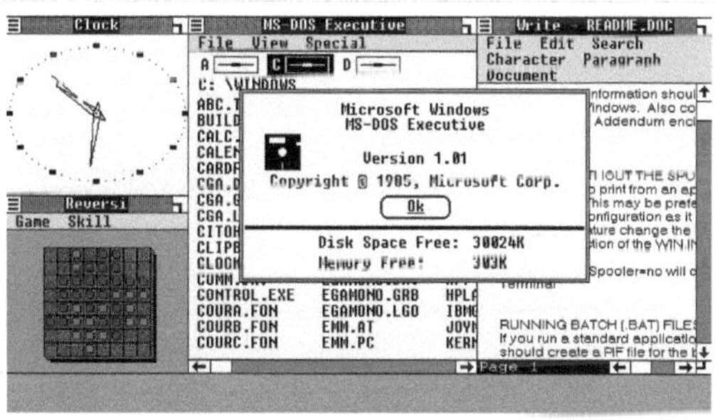

Windows 1.0, the first version, released in 1985

The history of Windows dates back to September 1981, when Chase Bishop, a computer scientist, designed the first model of an electronic device and project "Interface Manager" was started. It was announced in November 1983 (after the Apple Lisa, but before the Macintosh) under the name "Windows", but Windows 1.0 was not released until November 1985.[8] The shell of Windows 1.0 was a program known as the MS-DOS Executive. Other supplied programs were Calculator, Calendar, Cardfile, Clipboard viewer, Clock, Control Panel, Notepad, Paint, Reversi, Terminal, and Write. Windows 1.0 did not allow overlapping windows. Instead all windows were tiled. Only dialog boxes could appear over other windows.

Windows 2.0 was released in October 1987 and featured several improvements to the user interface and memory management.[8] Windows 2.0 allowed application windows to overlap each other and also introduced more sophisticated keyboard shortcuts. It could also make use of expanded memory.

Windows 2.1 was released in two different versions: Windows/386 employed the 386 virtual 8086 mode to multitask several DOS programs, and the paged memory model to emulate expanded memory using available extended memory. Windows/286 (which, despite its name, would run on the 8086) still ran in real mode, but could make use of the high memory area.

The early versions of Windows were often thought of as simply graphical user interfaces, mostly because they ran on top of MS-DOS and used it for file system services.[9] However, even the earliest 16-bit Windows versions already assumed many typical operating system functions; notably, having their own executable file format and providing their own device drivers (timer, graphics, printer, mouse, keyboard and sound) for applications. Unlike MS-DOS, Windows allowed users to execute multiple graphical applications at the same time, through cooperative multitasking. Windows implemented an elaborate, segment-based, software virtual memory scheme, which allowed it to run applications larger than available memory: code segments and resources were swapped in and thrown away when memory became scarce, and data segments moved in memory when a given application had relinquished processor control.

Windows 3.0 and 3.1

Windows 3.0 (1990) and Windows 3.1 (1992) improved the design, mostly because of virtual memory and loadable virtual device drivers (VxDs) that allowed them to share arbitrary devices between multitasked DOS windows. Also, Windows applications could now run in protected mode (when Windows was running in Standard or 386 Enhanced Mode), which gave them access to several megabytes of memory and removed the obligation to participate in the software virtual memory scheme. They still ran inside the same address space, where the segmented memory provided a degree of protection, and multi-tasked cooperatively. For Windows 3.0, Microsoft also rewrote critical operations from C into assembly.

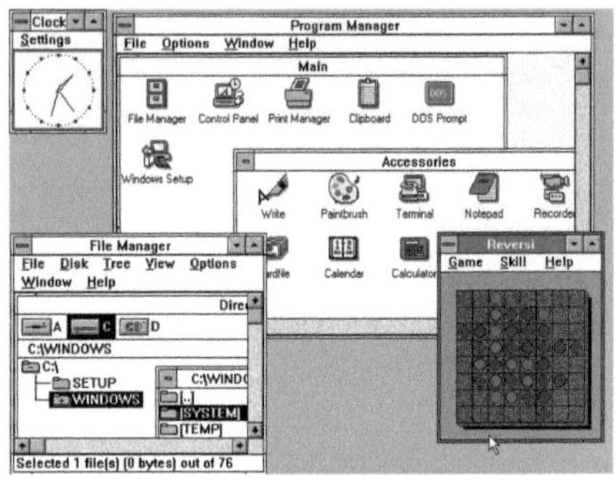

Windows 3.0, released in 1990

Windows 95, 98, and Me

Windows 95 was released in August 1995, featuring a new user interface, support for long file names of up to 255 characters, and the ability to automatically detect and configure installed hardware (plug and play). It could natively run 32-bit applications, and featured several technological improvements that increased its stability over Windows 3.1. There were several OEM Service Releases (OSR) of Windows 95, each of which was roughly equivalent to a service pack.

Microsoft's next release was Windows 98 in June 1998. Microsoft released a second version of Windows 98 in May 1999, named Windows 98 Second Edition (often shortened to Windows 98 SE).

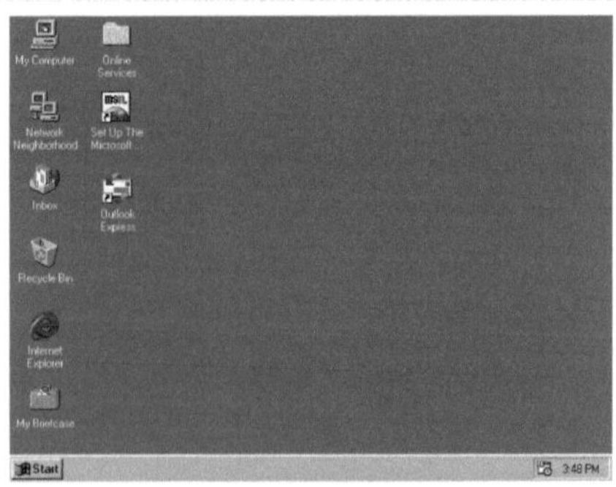

Windows 95, released in August 1995

In February 2000, Windows 2000 (in the NT family) was released, followed by Windows Me in September 2000 (*Me* standing for *Millennium Edition*). Windows Me updated the core from Windows 98, but adopted some aspects of Windows 2000 and removed the "boot in DOS mode" option. It also added a new feature called System Restore, allowing the user to set the computer's settings back to an earlier date.

Windows Millennium Edition is often confused with Windows 2000 (because of its name), and has been said to be one of the worst operating systems Microsoft ever released.[10]

Windows NT family

The NT family of Windows systems was fashioned and marketed for higher reliability business use. The first release was NT 3.1 (1993), numbered "3.1" to match the consumer Windows version, which was followed by NT 3.5 (1994), NT 3.51 (1995), NT 4.0 (1996), and Windows 2000, which is the last NT-based Windows release that does not include Microsoft Product Activation. Windows NT 4.0 was the first in this line to implement the "Windows 95" user interface (and the first to include Windows 95's built-in 32-bit runtimes).

Microsoft then moved to combine their consumer and business operating systems with Windows XP that was released on October 25, 2001. It came both in home and professional versions (and later niche market versions for tablet PCs and media centers); they also diverged release schedules for server operating systems. Windows Server 2003, released a year and a half after Windows XP, brought Windows Server up to date with Windows XP. After a lengthy development process, Windows Vista was released on November 30, 2006 for volume licensing and January 30, 2007 for consumers. And its server counterpart, Windows Server 2008 was released in early 2008. On July 22, 2009, Windows 7 and Windows Server 2008 R2 were released as RTM (release to manufacturing) while the former was released to the public 3 months later on October 22, 2009.

64-bit operating systems

Windows NT included support for several different platforms before the x86-based personal computer became dominant in the professional world. Versions of NT from 3.1 to 4.0 variously supported PowerPC, DEC Alpha and MIPS R4000, some of which were 64-bit processors, although the operating system treated them as 32-bit processors.

With the introduction of the Intel Itanium architecture (also known as IA-64), Microsoft released new versions of Windows to support it. Itanium versions of Windows XP and Windows Server 2003 were released at the same time as their mainstream x86 (32-bit) counterparts. On April 25, 2005, Microsoft released Windows XP Professional x64 Edition and Windows Server 2003 x64 Editions to support the x86-64 (or *x64* in Microsoft terminology) architecture. Microsoft dropped support for the Itanium version of Windows XP in 2005. Windows Vista was the first end-user version of Windows that Microsoft released simultaneously in x86 and x64 editions. Windows Vista does not support the Itanium architecture. The modern 64-bit Windows family comprises AMD64/Intel64 versions of Windows 7 and Windows Server 2008, in both Itanium and x64 editions. Windows Server 2008 R2 drops the 32 bit version, although Windows 7 does not.

Windows CE

Windows CE (officially known as *Windows Embedded Compact*), is an edition of Windows that runs on minimalistic computers, like satellite navigation systems and some mobile phones. Windows Embedded Compact is based on its own dedicated kernel, dubbed Windows CE kernel. Microsoft licenses Windows CE to OEMs and device makers. The OEMs and device makers can modify and create their own user interfaces and experiences, while Windows CE provides the technical foundation to do so.

Windows CE was used in the Dreamcast along with Sega's own proprietary OS for the console. Windows CE is the core from which Windows Mobile is derived.

The latest current version of Windows CE, Windows Embedded Compact 7, displaying a possible UI for what the media player can look like.

Microsoft's latest mobile OS, Windows Phone, is based on components from both Windows CE 6.0 R3 and the current Windows CE 7.0.

Windows Embedded Compact is not to be confused with Windows XP Embedded or Windows NT 4.0 Embedded, modular editions of Windows based on Windows NT kernel.

Future of Windows

Screenshot of Windows 8 Consumer Preview startscreen (as of February 29th, 2012)

Windows 8, the successor to Windows 7, is currently in development. Microsoft posted a blog entry in Dutch on October 22, 2010 hinting that Windows 8 would be released in roughly two years.[11] Also, during the pre-Consumer Electronics Show keynote, Microsoft's CEO announced that Windows 8 will also run on ARM CPUs. This Windows version will also be more suitable for tablets and netbooks, featuring a more touch-friendly interface. Several new features will also be introduced, such as support for USB 3.0 and the ability to run Windows from USB devices (like USB Hard Disks or USB Flash drives) with Windows To Go.

Microsoft has also announced a version of Windows 8 for ARM devices, named Windows RT.[12]

Bootable Windows To Go USB flash drive

History

The first version of Microsoft Windows, version 1.0, released in November 1985, lacked a degree of functionality, achieved little popularity and was to compete with Apple's own operating system. Windows 1.0 is not a complete operating system; rather, it extends MS-DOS. Microsoft Windows version 2.0 was released in November 1987 and was slightly more popular than its predecessor. Windows 2.03 (release date January 1988) had changed the OS from tiled windows to overlapping windows. The result of this change led to Apple Computer filing a suit against Microsoft alleging infringement on Apple's copyrights.[13] [14]

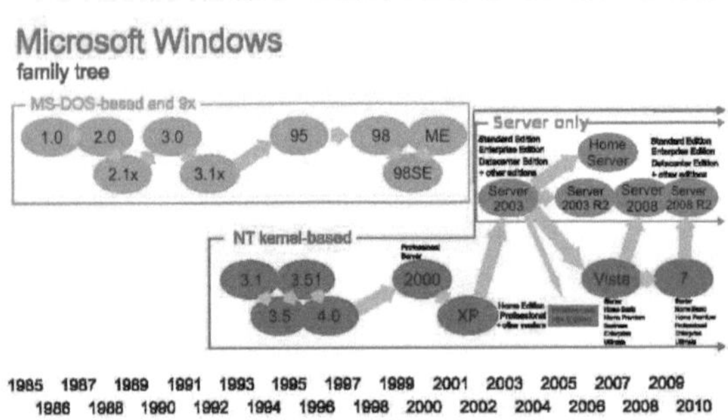

The Windows family tree.

Microsoft Windows version 3.0, released in 1990, was the first Microsoft Windows version to achieve broad commercial success, selling 2 million copies in the first six months.[15] [16] It featured improvements to the user interface and to multitasking capabilities. It received a facelift in Windows 3.1, made generally available on March 1, 1992. Windows 3.1 support ended on December 31, 2001.[17]

In July 1993, Microsoft released Windows NT based on a new kernel. Windows NT 3.1 was the first release of Windows NT. NT was considered to be the professional OS and was the first Windows version to utilize preemptive multitasking. Windows NT would later be retooled to also function as a home operating system, with Windows XP.

On August 24, 1995, Microsoft released Windows 95, a new, and major, consumer version that made further changes to the user interface, and also used preemptive multitasking. Windows 95 was designed to replace not only Windows 3.1, but also Windows for Workgroups, and MS-DOS. It was also the first Windows operating system to use Plug and Play capabilities. The changes Windows 95 brought to the desktop were revolutionary, as opposed to evolutionary, such as those in Windows 98 and Windows Me. Mainstream support for Windows 95 ended on December 31, 2000 and extended support for Windows 95 ended on December 31, 2001.[18]

Next in the consumer line was Microsoft Windows 98 released on June 25, 1998. It was followed with the release of Windows 98 Second Edition (Windows 98 SE) in 1999. Mainstream support for Windows 98 ended on June 30, 2002 and extended support for Windows 98 ended on July 11, 2006.[19]

As part of its "professional" line, Microsoft released Windows 2000 in February 2000. During 2004 part of the Source Code for Windows 2000 was leaked onto the Internet. This was bad for Microsoft as the same kernel used in Windows 2000 was used in Windows XP. The consumer version following Windows 98 was Windows Me (Windows Millennium Edition). Released in September 2000, Windows Me implemented a number of new technologies for Microsoft: most notably publicized was "Universal Plug and Play". Windows Me was heavily criticized due to slowness, freezes and hardware problems.

In October 2001, Microsoft released Windows XP, a version built on the Windows NT kernel that also retained the consumer-oriented usability of Windows 95 and its successors. This new version was widely praised in computer magazines.[20] It shipped in two distinct editions, "Home" and "Professional", the former lacking many of the superior security and networking features of the Professional edition. Additionally, the first "Media Center" edition was released in 2002,[21] with an emphasis on support for DVD and TV functionality including program recording and a remote control. Mainstream support for Windows XP ended on April 14, 2009. Extended support will continue until April 8, 2014.[22]

In April 2003, Windows Server 2003 was introduced, replacing the Windows 2000 line of server products with a number of new features and a strong focus on security; this was followed in December 2005 by Windows Server 2003 R2.

On January 30, 2007, Microsoft released Windows Vista. It contains a number of new features, from a redesigned shell and user interface to significant technical changes, with a particular focus on security features. It is available in a number of different editions, and has been subject to some criticism.

On October 22, 2009, Microsoft released Windows 7. Unlike its predecessor, Windows Vista, which introduced a large number of new features, Windows 7 was intended to be a more focused, incremental upgrade to the Windows line, with the goal of being compatible with applications and hardware which Windows Vista was not at the time.[23] Windows 7 has multi-touch support, a redesigned Windows shell with a new taskbar, referred to as the Superbar, a home networking system called HomeGroup,[24] and performance improvements.

On 29 February 2012, Microsoft released Windows 8 Consumer Preview, the beta version of Windows 8, build 8250. For the first time since Windows 95, the Start button is no longer available on the taskbar, though the Start screen is still triggered by clicking the bottom-left corner of the screen and by clicking Start in the Charm. Windows president Steven Sinofsky said more than 100,000 changes had been made since the developer version went public. In the first day of its release, Windows 8 Consumer Preview was downloaded over one million times. Like the Developer Preview, the Consumer Preview is set to expire on the 15 January 2013.

Timeline of releases

Usage share

Source	Net Market Share[26]	W3Counter[27]	Global Stats[28]
Date	March 2012	March 2012	March 2012
All versions	92.5%	76.7%	89.02%
Windows 7	37.54%	39.47%	46.48%
Windows XP	46.86%	29.38%	33.49%
Windows Vista	7.65%	7.77%	9.05%
Windows 2000	0.16%	—	—
Windows NT 4.0	0.12%	—	—
Windows 8	0.11%	0.08%	—
Windows 98	0.04%	—	—
Windows Me	0.01%	—	—
Windows 95	0.01%	—	—

Security

Consumer versions of Windows were originally designed for ease-of-use on a single-user PC without a network connection, and did not have security features built in from the outset.[29] However, Windows NT and its successors are designed for security (including on a network) and multi-user PCs, but were not initially designed with Internet security in mind as much, since, when it was first developed in the early 1990s, Internet use was less prevalent.[30]

These design issues combined with programming errors (e.g. buffer overflows) and the popularity of Windows means that it is a frequent target of computer worm and virus writers. In June 2005, Bruce Schneier's *Counterpane Internet Security* reported that it had seen over 1,000 new viruses and worms in the previous six months.[31] In 2005, Kaspersky Lab found around 11,000 malicious programs—viruses, Trojans, back-doors, and exploits written for Windows.[32]

Microsoft releases security patches through its Windows Update service approximately once a month (usually the second Tuesday of the month), although critical updates are made available at shorter intervals when necessary.[33] In versions of Windows after and including Windows 2000 SP3 and Windows XP, updates can be automatically downloaded and installed if the user selects to do so. As a result, Service Pack 2 for Windows XP, as well as Service Pack 1 for Windows Server 2003, were installed by users more quickly than it otherwise might have been.[34]

While the Windows 9x series offered the option of having profiles for multiple users, they had no concept of access privileges, and did not allow concurrent access; and so were not true multi-user operating systems. In addition, they implemented only partial memory protection. They were accordingly widely criticised for lack of security.

The Windows NT series of operating systems, by contrast, are true multi-user, and implement absolute memory protection. However, a lot of the advantages of being a true multi-user operating system were nullified by the fact that, prior to Windows Vista, the first user account created during the setup process was an administrator account, which was also the default for new accounts. Though Windows XP did have limited accounts, the majority of home users did not change to an account type with fewer rights – partially due to the number of programs which unnecessarily required administrator rights – and so most home users ran as administrator all the time.

Windows Vista changes this[35] by introducing a privilege elevation system called User Account Control. When logging in as a standard user, a logon session is created and a token containing only the most basic privileges is assigned. In this way, the new logon session is incapable of making changes that would affect the entire system. When logging in as a user in the Administrators group, two separate tokens are assigned. The first token contains all privileges typically awarded to an administrator, and the second is a restricted token similar to what a standard user would receive. User applications, including the Windows Shell, are then started with the restricted token, resulting in a reduced privilege environment even under an Administrator account. When an application requests higher privileges or "Run as administrator" is clicked, UAC will prompt for confirmation and, if consent is given (including administrator credentials if the account requesting the elevation is not a member of the administrators group), start the process using the unrestricted token.[36]

File permissions

All Windows versions from Windows NT 3 have been based on a file system permission system referred to as AGLP (Accounts, Global, Local, Permissions) AGDLP which in essence where file permissions are applied to the file/folder in the form of a 'local group' which then has other 'global groups' as members. These global groups then hold other groups or users depending on different Windows versions used. This system varies from other vendor products such as Linux and NetWare due to the 'static' allocation of permission being applied directory to the file or folder. However using this process of AGLP/AGDLP/AGUDLP allows a small number of static permissions to be applied and allows for easy changes to the account groups without reapplying the file permissions on the files and folders.

Windows Defender

On January 6, 2005, Microsoft released a Beta version of Microsoft AntiSpyware, based upon the previously released Giant AntiSpyware. On February 14, 2006, Microsoft AntiSpyware became Windows Defender with the release of Beta 2. Windows Defender is a freeware program designed to protect against spyware and other unwanted software. Windows XP and Windows Server 2003 users who have genuine copies of Microsoft Windows can freely download the program from Microsoft's web site, and Windows Defender ships as part of Windows Vista and 7.[37] In Windows 8, Windows Defender and Microsoft Security Essentials have been combined into a single program, named Windows Defender.It is based on Microsoft Security Essentials borrowing its features and user interface. Although it is enabled by default, it can be turned off to use another anti-virus solution.[38]

Third-party analysis

In an article based on a report by Symantec,[39] internetnews.com has described Microsoft Windows as having the "fewest number of patches and the shortest average patch development time of the five operating systems it monitored in the last six months of 2006."[40]

A study conducted by Kevin Mitnick and marketing communications firm Avantgarde in 2004 found that an unprotected and unpatched Windows XP system with Service Pack 1 lasted only 4 minutes on the Internet before it was compromised, and an unprotected and also unpatched Windows Server 2003 system was compromised after being connected to the internet for 8 hours.[41] This study does not apply to Windows XP systems running the Service Pack 2 update (released in late 2004), which vastly improved the security of Windows XP. The computer that was running Windows XP Service Pack 2 was not compromised. The AOL National Cyber Security Alliance Online Safety Study of October 2004 determined that 80% of Windows users were infected by at least one spyware/adware product. Much documentation is available describing how to increase the security of Microsoft Windows products. Typical suggestions include deploying Microsoft Windows behind a hardware or software firewall, running anti-virus and anti-spyware software, and installing patches as they become available through Windows Update.[42]

Emulation software

Emulation allows the use of some Windows applications without using Microsoft Windows. These include:

- Wine – a free and open source software implementation of the Windows API, allowing one to run many Windows applications on x86-based platforms, including Linux and Mac OS X. Wine developers refer to it as a "compatibility layer";[43] and make use of Windows-style APIs to emulate the Windows environment.
    - CrossOver – A Wine package with licensed fonts. Its developers are regular contributors to Wine, and focus on Wine running officially supported applications.
    - Cedega – TransGaming Technologies' proprietary fork of Wine, designed specifically for running games written for Microsoft Windows under Linux. A version of Cedega known as Cider is used by some video game publishers to allow Windows games to run on Mac OS X. Since Wine was licensed under the LGPL, Cedega has been unable to port the improvements made to Wine to their proprietary codebase. Cedega ceased its service in February 2011.
    - Darwine – A bundling of Wine to the PowerPC Macs running OS X by running Wine on top of QEMU. Intel Macs use the same Wine as other *NIX x86 systems.
- ReactOS – An open-source OS that is intended to run the same software as Windows, originally designed to simulate Windows NT 4.0, now aiming at Windows XP and Vista/7 compatibility. It has been in the development stage since 1996.

See also

- Architecture of the Windows NT operating system line
- Criticism of Microsoft Windows
- Comparison of operating systems
- Comparison of Windows versions
- List of Microsoft Windows components
- List of operating systems
- Wintel

References

[1] "NT Server Training: Architectural Overview. Lesson 2 – Windows NT System Overview." (http://www.microsoft.com/technet/archive/winntas/training/ntarchitectoview/ntarc_2.mspx). *Microsoft TechNet*. Microsoft. . Retrieved December 9, 2010.

[2] http://en.wikipedia.org/wiki/Template%3Alatest_stable_software_release%2Fwindows_nt

[3] http://en.wikipedia.org/wiki/Template%3Alatest_stable_software_release%2Fmicrosoft_windows

[4] http://en.wikipedia.org/wiki/Template%3Alatest_preview_software_release%2Fwindows_nt

[5] http://en.wikipedia.org/wiki/Template%3Alatest_preview_software_release%2Fmicrosoft_windows

[6] http://msdn.microsoft.com/goglobal/ee461121#AvailableLanguagePacks

[7] "The Unusual History of Microsoft Windows" (http://inventors.about.com/od/mstartinventions/a/Windows.htm?rd=1). . Retrieved April 22, 2007.

[8] Petzold

[9] "Windows Evolution" (http://news.soft32.com/windows-evolution_1629.html). Soft32.com News. .

[10] "The 25 Worst Tech Products of All Time" (http://www.pcworld.com/article/125772-2/the_25_worst_tech_products_of_all_time.html). PC World. . Retrieved February 10, 2012.

[11] "Microsoft says Windows 8 roughly two years away" (http://news.cnet.com/8301-13860_3-20020544-56.html?tag=mncol;title). *CNET News* (CNET). October 24, 2010. . Retrieved December 9, 2010.

[12] "Windows Announcing the Windows 8 Editions" (http://windowsteamblog.com/windows/b/bloggingwindows/archive/2012/04/16/announcing-the-windows-8-editions.aspx). The Windows Blog. .

[13] "The Apple vs. Microsoft GUI Lawsuit" (http://lowendmac.com/orchard/06/apple-vs-microsoft.html). 2006. . Retrieved March 12, 2008.

[14] "Apple Computer, Inc. v. MicroSoft Corp., 35 F.3d 1435 (9th Cir. 1994)" (http://home.earthlink.net/~mjohnsen/Technology/Lawsuits/appvsms.html). . Retrieved March 12, 2008.

[15] "Chronology of Personal Computer Software" (http://www.islandnet.com/~kpolsson/compsoft/soft1991.htm). .

[16] "Microsoft Company" (http://www.thocp.net/companies/microsoft/microsoft_company.htm). .

[17] "Windows 3.1 Standard Edition Support Lifecycle" (http://support.microsoft.com/lifecycle/?LN=en-us&p1=3078&x=10&y=11). . Retrieved January 3, 2011.

[18] "Windows 95 Support Lifecycle" (http://support.microsoft.com/lifecycle/?p1=7864). Microsoft. . Retrieved January 3, 2011.

[19] "Windows 98 Standard Edition Support Lifecycle" (http://support.microsoft.com/lifecycle/?p1=6513). Microsoft. . Retrieved January 3, 2011.

[20] David Coursey (August 31, 2001). "Your top Windows XP questions answered! (Part One)" (http://web.archive.org/web/20071219121319/http://review.zdnet.com/4520-6033_16-4206367.html). *ZDNet*. CNET. Archived from the original (http://review.zdnet.com/4520-6033_16-4206367.html) on December 19, 2007. . Retrieved January 3, 2011.

[21] "A Look at Freestyle and Mira" (http://www.winsupersite.com/article/showcase/a-look-at-freestyle-and-mira.aspx). *Paul Thurrott's SuperSite for Windows*. Penton. September 3, 2002. . Retrieved January 3, 2011.

[22] "Windows XP Professional Lifecycle Support" (http://support.microsoft.com/lifecycle/?p1=3223). . Retrieved January 3, 2011.

[23] Mike Nash (October 28, 2008). "Windows 7 Unveiled Today at PDC 2008" (http://windowsteamblog.com/blogs/windows7/archive/2008/10/28/windows-7-unveiled-today-at-pdc-2008.aspx). *Windows Team Blog*. Microsoft. . Retrieved November 11, 2008.

[24] Brandon LeBlanc (October 28, 2008). "How Libraries & HomeGroup Work Together in Windows 7" (http://windowsteamblog.com/blogs/windowsexperience/archive/2008/10/28/how-libraries-amp-homegroup-work-together-in-windows-7.aspx). *Windows Team Blog*. Microsoft. . Retrieved November 11, 2008.

[25] "Windows 98 Second Edition Support Lifecycle" (http://support.microsoft.com/lifecycle/?p1=6898). Microsoft. . Retrieved January 3, 2011.

[26] "Operating System Market Share" (http://marketshare.hitslink.com/operating-system-market-share.aspx?qprid=10&qpcustomd=0&qptimeframe=M&qpsp=158&qpnp=1). *Net Market Share*. Net Applications. March 2012. . Retrieved April 14, 2012.

[27] "Global Web Stats" (http://www.w3counter.com/globalstats.php?year=2012&month=3). *W3Counter*. Awio Web Services. March 2012. . Retrieved April 14, 2012.

[28] "StatCounter Global Stats" (http://gs.statcounter.com/#os-ww-monthly-201203-201203-bar). *Global Stats*. StatCounter. March 2012. . Retrieved April 14, 2012.

[29] Multi-user memory protection was not introduced until Windows NT and XP, and a computer's default user was an administrator until Windows Vista. Source: UACBlog (http://blogs.msdn.com/uac/).

[30] "Telephones and Internet Users by Country, 1990 and 2005" (http://www.infoplease.com/ipa/A0883396.html). Information Please Database. . Retrieved June 9, 2009.

[31] Bruce Schneier (June 15, 2005). "Crypto-Gram Newsletter" (http://www.schneier.com/crypto-gram-0506.html). Counterpane Internet Security, Inc.. . Retrieved April 22, 2007.

[32] Andy Patrizio (April 27, 2006). "Linux Malware On The Rise" (http://www.internetnews.com/dev-news/article.php/3601946). *InternetNews*. QuinStreet. . Retrieved January 3, 2011.

[33] Ryan Naraine (June 8, 2005). "Microsoft's Security Response Center: How Little Patches Are Made" (http://www.eweek.com/c/a/Windows/Microsofts-Security-Response-Center-How-Little-Patches-Are-Made/). *eWeek*. Ziff Davis Enterprise. . Retrieved January 3, 2011.

[34] John Foley (October 20, 2004). "Windows XP SP2 Distribution Surpasses 100 Million" (http://www.informationweek.com/news/security/vulnerabilities/showArticle.jhtml?articleID=50900297). *InformationWeek*. UBM TechWeb. . Retrieved January 3, 2011.

[35] Microsoft describes in detail the steps taken to combat this in a TechNet bulletin. Source: Windows Vista Security and Data Protection Improvements (http://technet.microsoft.com/en-us/windowsvista/aa905073.aspx).

[36] Kenny Kerr (September 29, 2006). "Windows Vista for Developers – Part 4 – User Account Control" (http://weblogs.asp.net/kennykerr/archive/2006/09/29/Windows-Vista-for-Developers-_1320_-Part-4-_1320_-User-Account-Control.aspx). . Retrieved March 15, 2007.

[37] "Windows Vista: Security & Safety" (http://windows.microsoft.com/en-US/windows-vista/products/features/security-safety). Microsoft. . Retrieved April, 16 2012.

[38] "Microsoft Answers: How do I keep Windows 8 Consumer Preview secure from malware?" (http://answers.microsoft.com/en-us/windows/forum/windows_8-security/how-do-i-keep-windows-8-consumer-preview-secure/acd2bfea-ed36-401e-9050-f2fe4212ecf3). Microsoft. . Retrieved April 16, 2012.

[39] "Symantec Internet Security Threat Report Trends for July – December 2006" (http://eval.symantec.com/mktginfo/enterprise/white_papers/ent-whitepaper_internet_security_threat_report_xi_03_2007.en-us.pdf) (PDF). *Internet Security Threat Report Volume XI*. Symantec. March 2007. . Retrieved January 3, 2011.

[40] Andy Patrizio (March 21, 2007). "Report Says Windows Gets The Fastest Repairs" (http://www.internetnews.com/security/article.php/3667201). *InternetNews*. QuinStreet. . Retrieved January 3, 2011.

[41] "Automated "Bots" Overtake PCs Without Firewalls Within 4 Minutes" (http://www.avantgarde.com/ttln113004.html). *Avantgarde*. Avantgarde. November 30, 2004. . Retrieved January 3, 2011.

[42] Richard Rogers (September 21, 2009). "5 Steps To Securing Your Windows XP Home Computer" (http://www.computer-security-news.com/0969/5-steps-to-securing-your-windows-xp-home-computer). *Computer Security News*. Computer Security News. . Retrieved January 3, 2011.

[43] Wine (http://www.winehq.org/)

External links

- Official website (http://windows.microsoft.com/en-US/windows/home)
- Microsoft Developer Network (http://www.msdn.com/)
- Windows Client Developer Resources (http://msdn.microsoft.com/en-au/windows/default.aspx?WT.mc_id=soc-c-au-loc--2010oct)
- Microsoft Windows History Timeline (http://www.microsoft.com/Windows/WinHistoryIntro.mspx)
- Pearson Education, InformIT (http://www.informit.com/articles/article.aspx?p=1358665&rll=1) – History of Microsoft Windows
- Microsoft Windows 7 for Government (http://www.microsoft.com/industry/government/products/windows7/default.aspx)

UDP_Data_Transport

UDT (acronym for **UDP Data Transport**) is a protocol developed by the National Center for Data Mining (NCDM) at University of Illinois at Chicago. The protocol was designed to sit on top of User Datagram Protocol (UDP) to facilitate large data transfers over modern high-speed fiber optic computer networks. In October, 2003 the NCDM achieved a 6.8 gigabits per second transfer from Chicago, USA to Amsterdam, Netherlands. During the 30-minute test they transmitted approximately 1.4 terabytes of data.

External links

- National Center for Data Mining [1]
- Sourceforge Project [2]
- IETF Draft from December 2003 [3]
- Third-party implementations [4]

References

[1] http://www.ncdm.uic.edu/
[2] http://udt.sourceforge.net/
[3] http://www.potaroo.net/ietf/all-ids/draft-gg-udt-00.txt
[4] http://udt.sourceforge.net/poweredby.html

User_Datagram_Protocol

The **User Datagram Protocol (UDP)** is one of the core members of the Internet Protocol Suite, the set of network protocols used for the Internet. With UDP, computer applications can send messages, in this case referred to as *datagrams*, to other hosts on an Internet Protocol (IP) network without requiring prior communications to set up special transmission channels or data paths. The protocol was designed by David P. Reed in 1980 and formally defined in RFC 768.

UDP uses a simple transmission model without implicit handshaking dialogues for providing reliability, ordering, or data integrity. Thus, UDP provides an unreliable service and datagrams may arrive out of order, appear duplicated, or go missing without notice. UDP assumes that error checking and correction is either not necessary or performed in the application, avoiding the overhead of such processing at the network interface level. Time-sensitive applications often use UDP because dropping packets is preferable to waiting for delayed packets, which may not be an option in a real-time system.[1] If error correction facilities are needed at the network interface level, an application may use the Transmission Control Protocol (TCP) or Stream Control Transmission Protocol (SCTP) which are designed for this purpose.

UDP's stateless nature is also useful for servers answering small queries from huge numbers of clients. Domain Name System (DNS), streaming media applications such as IPTV, Voice over IP (VoIP), Trivial File Transfer Protocol (TFTP), IP tunneling protocols and many online games.

Service ports

UDP applications use datagram sockets to establish host-to-host communications. An application binds a socket to its endpoint of data transmission, which is a combination of an IP address and a service port. A port is a software structure that is identified by the port number, a 16 bit integer value, allowing for port numbers between 0 and 65535. Port 0 is reserved, but is a permissible source port value if the sending process does not expect messages in response.

The Internet Assigned Numbers Authority has divided port numbers into three ranges.[2] Port numbers 0 through 1023 are used for common, well-known services. On Unix-like operating systems, using one of these ports requires superuser operating permission. Port numbers 1024 through 49151 are the registered ports used for IANA-registered services. Ports 49152 through 65535 are dynamic ports that are not officially designated for any specific service, and can be used for any purpose. They are also used as ephemeral ports, from which software running on the host may randomly choose a port in order to define itself.[2] In effect, they are used as temporary ports primarily by clients when communicating with servers.

Packet structure

UDP is a minimal message-oriented Transport Layer protocol that is documented in IETF RFC 768.

UDP provides no guarantees to the upper layer protocol for message delivery and the UDP protocol layer retains no state of UDP messages once sent. For this reason, UDP is sometimes referred to as *Unreliable* Datagram Protocol.[3]

UDP provides application multiplexing (via port numbers) and integrity verification (via checksum) of the header and payload.[4] If transmission reliability is desired, it must be implemented in the user's application.

offset (bits)	0 – 15	16 – 31
0	Source Port Number	Destination Port Number
32	Length	Checksum
64+	Data	

The UDP header consists of 4 fields, each of which is 2 bytes (16 bits).[1] The use of two of those is optional in IPv4 (pink background in table). In IPv6 only the source port is optional (see below).

Source port number

This field identifies the sender's port when meaningful and should be assumed to be the port to reply to if needed. If not used, then it should be zero. If the source host is the client, the port number is likely to be an ephemeral port number. If the source host is the server, the port number is likely to be a well-known port number.[2]

Destination port number

This field identifies the receiver's port and is required. Similar to source port number, if the client is the destination host then the port number will likely be an ephemeral port number and if the destination host is the server then the port number will likely be a well-known port number.[2]

Length

A field that specifies the length in bytes of the entire datagram: header and data. The minimum length is 8 bytes since that's the length of the header. The field size sets a theoretical limit of 65,535 bytes (8 byte header + 65,527 bytes of data) for a UDP datagram. The practical limit for the data length which is imposed by the underlying IPv4 protocol is 65,507 bytes (65,535 – 8 byte UDP header – 20 byte IP header).[2]

In IPv6 Jumbograms it is possible to have UDP packets of size greater than 65,535 bytes.[5] This allows for a maximum length value of 4,294,967,295 bytes (2^32 - 1) with 8 bytes representing the header and 4,294,967,287 bytes for data.

Checksum

The checksum field is used for error-checking of the header *and* data. If no checksum is generated by the transmitter, the field uses the value all-zeros.[6] This field is not optional for IPv6.[7]

Checksum computation

The method used to compute the checksum is defined in RFC 768:

Checksum is the 16-bit one's complement of the one's complement sum of a pseudo header of information from the IP header, the UDP header, and the data, padded with zero octets at the end (if necessary) to make a multiple of two octets.[6]

In other words, all 16-bit words are summed using one's complement arithmetic. The sum is then one's complemented to yield the value of the UDP checksum field.

If the checksum calculation results in the value zero (all 16 bits 0) it should be sent as the one's complement (all 1s).

The difference between IPv4 and IPv6 is in the data used to compute the checksum.

IPv4 PSEUDO-HEADER

When UDP runs over IPv4, the checksum is computed using a PSEUDO-HEADER that contains some of the same information from the real IPv4 header. The PSEUDO-HEADER is not the real IPv4 header used to send an IP packet. The following table defines the PSEUDO-HEADER used only for the checksum calculation.

bits	0 – 7	8 – 15	16 – 23	24 – 31
0	Source address			
32	Destination address			
64	Zeros	Protocol	UDP length	
96	Source Port		Destination Port	
128	Length		Checksum	
160+	Data			

The source and destination addresses are those in the IPv4 header. The protocol is that for UDP (see *List of IP protocol numbers*): 17 (0x11). The UDP length field is the length of the UDP header and data.

UDP checksum computation is optional for IPv4. If a checksum is not used it should be set to the value zero.

IPv6 PSEUDO-HEADER

When UDP runs over IPv6, the checksum is mandatory. The method used to compute it is changed as documented in RFC 2460:

> *Any transport or other upper-layer protocol that includes the addresses from the IP header in its checksum computation must be modified for use over IPv6 to include the 128-bit IPv6 addresses.*[7]

When computing the checksum, again a PSEUDO-HEADER is used that mimics the real IPv6 header:

bits	0 – 7	8 – 15	16 – 23	24 – 31
0	Source address			
32				
64				
96				
128	Destination address			
160				
192				
224				
256	UDP length			
288	Zeros			Next Header
320	Source Port		Destination Port	
352	Length		Checksum	
384+	Data			

The source address is the one in the IPv6 header. The destination address is the final destination; if the IPv6 packet does not contain a Routing header, that will be the destination address in the IPv6 header; otherwise, at the originating node, it will be the address in the last element of the Routing header, and, at the receiving node, it will be the destination address in the IPv6 header. The value of the Next Header field is the protocol value for UDP: 17. The UDP length field is the length of the UDP header and data.

Reliability and congestion control solutions

Lacking reliability, UDP applications must generally be willing to accept some loss, errors or duplication. Some applications such as TFTP may add rudimentary reliability mechanisms into the application layer as needed.[2]

Most often, UDP applications do not employ reliability mechanisms and may even be hindered by them. Streaming media, real-time multiplayer games and voice over IP (VoIP) are examples of applications that often use UDP. In these particular applications, loss of packets is not usually a fatal problem. If an application requires a high degree of reliability, a protocol such as the Transmission Control Protocol or erasure codes may be used instead.

Potentially more seriously, unlike TCP, UDP-based applications don't necessarily have good congestion avoidance and control mechanisms. Congestion insensitive UDP applications that consume a large fraction of available bandwidth could endanger the stability of the internet, as they frequently give a bandwidth load that is inelastic. Network-based mechanisms have been proposed to minimize potential congestion collapse effects of uncontrolled, high rate UDP traffic loads. Network-based elements such as routers using packet queuing and dropping techniques are often the only tool available to slow down excessive UDP traffic. The Datagram Congestion Control Protocol (DCCP) is being designed as a partial solution to this potential problem by adding end host TCP-friendly congestion control behavior to high-rate UDP streams such as streaming media.

Applications

Numerous key Internet applications use UDP, including: the Domain Name System (DNS), where queries must be fast and only consist of a single request followed by a single reply packet, the Simple Network Management Protocol (SNMP), the Routing Information Protocol (RIP)[1] and the Dynamic Host Configuration Protocol (DHCP).

Voice and video traffic is generally transmitted using UDP. Real-time video and audio streaming protocols are designed to handle occasional lost packets, so only slight degradation in quality occurs, rather than large delays if lost packets were retransmitted. Because both TCP and UDP run over the same network, many businesses are finding that a recent increase in UDP traffic from these real-time applications is hindering the performance of applications using TCP, such as point of sale, accounting, and database systems. When TCP detects packet loss, it will throttle back its data rate usage. Since both real-time and business applications are important to businesses, developing quality of service solutions is seen as crucial by some.[8]

Comparison of UDP and TCP

Transmission Control Protocol is a connection-oriented protocol, which means that it requires handshaking to set up end-to-end communications. Once a connection is set up user data may be sent bi-directionally over the connection.

- *Reliable* – TCP manages message acknowledgment, retransmission and timeout. Multiple attempts to deliver the message are made. If it gets lost along the way, the server will re-request the lost part. In TCP, there's either no missing data, or, in case of multiple timeouts, the connection is dropped.
- *Ordered* – if two messages are sent over a connection in sequence, the first message will reach the receiving application first. When data segments arrive in the wrong order, TCP buffers the out-of-order data until all data can be properly re-ordered and delivered to the application.
- *Heavyweight* – TCP requires three packets to set up a socket connection, before any user data can be sent. TCP handles reliability and congestion control.
- *Streaming* – Data is read as a byte stream, no distinguishing indications are transmitted to signal message (segment) boundaries.

UDP is a simpler message-based connectionless protocol. Connectionless protocols do not set up a dedicated end-to-end connection. Communication is achieved by transmitting information in one direction from source to destination without verifying the readiness or state of the receiver. However, one primary benefit of UDP over TCP is the application to voice over internet protocol (VoIP) where any handshaking would hinder clear voice communication. It is assumed in VoIP UDP that the end users provide any necessary real time confirmation that the message has been received.

- *Unreliable* – When a message is sent, it cannot be known if it will reach its destination; it could get lost along the way. There is no concept of acknowledgment, retransmission or timeout.
- *Not ordered* – If two messages are sent to the same recipient, the order in which they arrive cannot be predicted.
- *Lightweight* – There is no ordering of messages, no tracking connections, etc. It is a small transport layer designed on top of IP.
- *Datagrams* – Packets are sent individually and are checked for integrity only if they arrive. Packets have definite boundaries which are honored upon receipt, meaning a read operation at the receiver socket will yield an entire message as it was originally sent.
- *No congestion control* – UDP itself does not avoid congestion, and it's possible for high bandwidth applications to trigger congestion collapse, unless they implement congestion control measures at the application level.

See also

- List of TCP and UDP port numbers
- Reliable User Datagram Protocol (RUDP)
- SCTP
- Transport protocol comparison table
- UDP flood attack
- UDP Data Transport
- UDP Lite, a variant that will deliver packets even if they are malformed
- UDP Helper Address

References

[1] Kurose, J. F.; Ross, K. W. (2010). *Computer Networking: A Top-Down Approach* (5th ed.). Boston, MA: Pearson Education. ISBN 9780131365483.

[2] Forouzan, B.A. (2000). *TCP/IP: Protocol Suite, 1st ed.* New Delhi, India: Tata McGraw-Hill Publishing Company Limited.

[3] content@ipv6.com. "UDP Protocol Overview" (http://ipv6.com/articles/general/User-Datagram-Protocol.htm). Ipv6.com. . Retrieved 17 August 2011.

[4] Clark, M.P. (2003). *Data Networks IP and the Internet, 1st ed.* West Sussex, England: John Wiley & Sons Ltd.

[5] RFC 2675

[6] Postel, J. (August 1980). RFC 768: User Datagram Protocol. *Internet Engineering Task Force*. Retrieved from http://tools.ietf.org/html/rfc768

[7] Deering S. & Hinden R. (December 1998). RFC 2460: Internet Protocol, Version 6 (IPv6) Specification. *Internet Engineering Task Force*. Retrieved from http://tools.ietf.org/html/rfc2460

[8] "The impact of UDP on Data Applications" (http://www.networkperformancedaily.com/2007/08/whiteboard_series_nice_guys_fi.html). Networkperformancedaily.com. . Retrieved 17 August 2011.

RFC references

- RFC 768 – User Datagram Protocol
- RFC 2460 – Internet Protocol, Version 6 (IPv6) Specification
- RFC 2675 - IPv6 Jumbograms
- RFC 4113 – Management Information Base for the UDP
- RFC 5405 – Unicast UDP Usage Guidelines for Application Designers

External links

- IANA Port Assignments (http://www.iana.org/assignments/port-numbers)
- The Trouble with UDP Scanning (PDF) (http://condor.depaul.edu/~jkristof/papers/udpscanning.pdf)
- Breakdown of UDP frame (http://www.networksorcery.com/enp/protocol/udp.htm)
- UDP on MSDN Magazine Sockets and WCF (http://msdn.microsoft.com/en-us/magazine/cc163648.aspx)
- UDP connections (http://www.faqs.org/docs/iptables/udpconnections.html)

Active_Directory

Active Directory (AD) is a directory service created by Microsoft for Windows domain networks. It is included in most Windows Server operating systems. Server computers that run Active Directory are called domain controllers.

Active Directory provides a central location for network administration and security. It authenticates and authorizes all users and computers in a Windows domain type network—assigning and enforcing security policies for all computers and installing or updating software. For example, when a user logs into a computer that is part of a Windows domain, Active Directory verifies the password and specifies whether the user is a system administrator or normal user.[1]

Active Directory uses Lightweight Directory Access Protocol (LDAP) versions 2 and 3, Kerberos and DNS.

History

Active Directory was previewed in 1999, released first with Windows 2000 Server edition, and revised to extend functionality and improve administration in Windows Server 2003. Additional improvements were made in Windows Server 2003 R2, Windows Server 2008 and Windows Server 2008 R2, and with the release of the latter the domain controller role was renamed **Active Directory Domain Services**.

Structure

Objects

An Active Directory structure is a hierarchical arrangement of information about objects. The objects fall into two broad categories: resources (e.g., printers) and security principals (user or computer accounts and groups). Security principals are assigned unique security identifiers (SIDs).

Each object represents a single entity—whether a user, a computer, a printer, or a group—and its attributes. Certain objects can contain other objects. An object is uniquely identified by its name and has a set of attributes—the characteristics and information that the object represents— defined by a schema, which also determines the kinds of objects that can be stored in Active Directory.

The schema object lets administrators extend or modify the schema when necessary. However, because each schema object is integral to the definition of Active Directory objects, deactivating or changing these objects can fundamentally change or disrupt a deployment. Schema changes automatically propagate throughout the system. Once created, an object can only be deactivated—not deleted. Changing the schema usually requires planning.[2]

Sites

A **Site** object in Active Directory represents a geographic location that hosts networks.[3]

Forests, trees, and domains

The Active Directory framework that holds the objects can be viewed at a number of levels. The forest, tree, and domain are the logical divisions in an Active Directory network.

Within a deployment, objects are grouped into domains. The objects for a single domain are stored in a single database (which can be replicated). Domains are identified by their DNS name structure, the namespace.

A tree is a collection of one or more domains and domain trees in a contiguous namespace, linked in a transitive trust hierarchy.

At the top of the structure is the *forest.* A forest is a collection of trees that share a common global catalog, directory schema, logical structure, and directory configuration. The forest represents the security boundary within which users, computers, groups, and other objects are accessible.

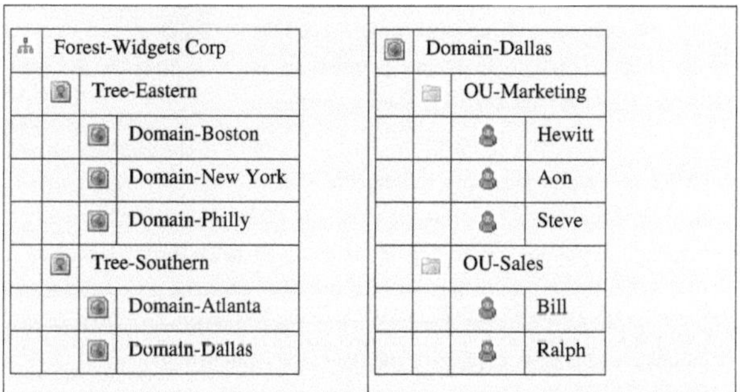

Example of the geographical organizing of zones of interest within trees and domains.

Organizational units

The objects held within a domain can be grouped into Organizational Units (OUs).[4] OUs can provide hierarchy to a domain, ease its administration, and can resemble the organization's structure in managerial or geographical terms. OUs can contain other OUs—domains are containers in this sense. Microsoft recommends using OUs rather than domains for structure and to simplify the implementation of policies and administration. The OU is the recommended level at which to apply group policies, which are Active Directory objects formally named Group Policy Objects (GPOs), although policies can also be applied to domains or sites (see below). The OU is the level at which administrative powers are commonly delegated, but delegation can be performed on individual objects or attributes as well.

Organizational Units are an abstraction for the administrator and do not function as containers; the underlying domain is the true container. It is not possible, for example, to create user accounts with an identical username (sAMAccountName) in separate OUs, such as "fred.staff-ou.domain" and "fred.student-ou.domain", where "staff-ou" and "student-ou" are the OUs. This is so because sAMAccountName, a user object attribute, must be unique within the domain. However, two users in different OUs can have the same Common Name (CN), the first component of the Distinguished Name (DN) of the user. Thus from the point of view of the DN, OUs do function as containers.

As the number of users in a domain increases, conventions such as "first initial, middle initial, last name" (Western order) or the reverse (Eastern order) fail for common family names like *Li* (), *Smith* or *Garcia.* Workarounds include adding a digit to the end of the username. Alternatives include creating a separate ID system of unique employee/student id numbers to use as account names in place of actual user's names, and allowing users to nominate their preferred word sequence within an acceptable use policy.

Because duplicate usernames cannot exist within a domain, account name generation poses a significant challenge for large organizations that cannot be easily subdivided into separate domains, such as students in a public school system or university who must be able to use any computer across the network.

Shadow groups

In Microsoft's Active Directory, OUs do not confer access permissions, and objects placed within OUs are not automatically assigned access privileges based on their containing OU. This is a design limitation specific to Active Directory. Other competing directories such as Novell NDS are able to assign access privileges through object placement within an OU.

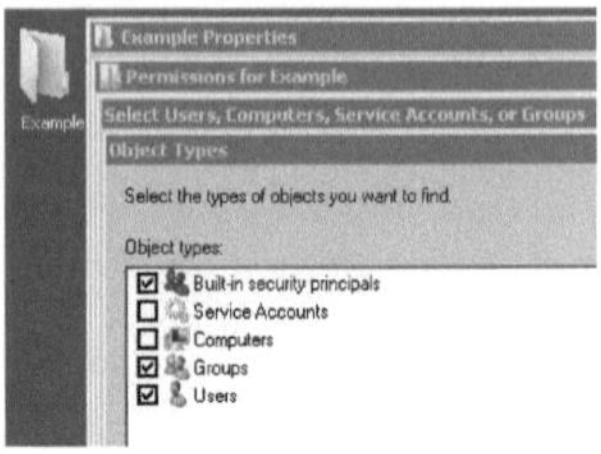

In Active Directory, organizational units cannot be assigned as owners or trustees. Only groups are selectable, and members of OUs cannot be collectively assigned rights to directory objects.

Active Directory requires a separate step for an administrator to assign an object in an OU as a member of a group also within that OU. Relying on OU location alone to determine access permissions is unreliable, because the object may not have been assigned to the group object for that OU.

A common workaround for an Active Directory administrator is to write a custom PowerShell or Visual Basic script to automatically create and maintain a *user group* for each OU in their directory. The scripts are run periodically to update the group to match the OU's account membership, but are unable to instantly update the security groups anytime the directory changes, as occurs in competing directories where security is directly implemented into the directory itself. Such groups are known as *Shadow Groups*. Once created, these shadow groups are selectable in place of the OU in the administrative tools.

Microsoft refers to shadow groups in the Server 2008 Reference documentation, but does not explain how to create them. There are no built-in server methods or console snap-ins for managing shadow groups.[5]

The division of an organization's information infrastructure into a hierarchy of one or more domains and top-level OUs is a key decision. Common models are by business unit, by geographical location, by IT Service, or by object type and hybrids of these. OUs should be structured primarily to facilitate administrative delegation, and secondarily, to facilitate group policy application. Although OUs form an administrative boundary, the only true security boundary is the forest itself and an administrator of any domain in the forest must be trusted across all domains in the forest.[6]

Physical matters

Sites are physical (rather than logical) groupings defined by one or more IP subnets.[7] AD also holds the definitions of connections, distinguishing low-speed (e.g., WAN, VPN) from high-speed (e.g., LAN) links. Site definitions are independent of the domain and OU structure and are common across the forest. Sites are used to control network traffic generated by replication and also to refer clients to the nearest domain controllers. Microsoft Exchange Server 2007 uses the site topology for mail routing. Policies can also be defined at the site level.

Physically the Active Directory information is held on one or more peer domain controllers (DCs), replacing the NT PDC/BDC model. Each DC has a copy of the Active Directory. Servers joined to Active Directory that are not domain controllers are called Member Servers.[8]

The Active Directory database is organized in *partitions*, each holding specific object types and following a specific replication pattern. AD synchronizes changes using *multi-master replication*.[9] Microsoft often refers to these partitions as 'naming contexts'.[10] The 'Schema' partition contains the definition of object classes and attributes within the Forest. The 'Configuration' partition contains information on the physical structure and configuration of the forest (such as the site topology). Both replicate to all domain controllers in the Forest. The 'Domain' partition holds all objects created in that domain and replicates only to Domain Controllers within its domain. So, for example, a user created in Domain X would be listed only in Domain X's domain controllers. A subset of objects in the domain partition replicate to domain controllers that are configured as global catalogs. Global catalog (GC) servers provide a global listing of all objects in the Forest.[11] Global Catalog servers replicate to themselves all objects from all domains and hence, provide a global listing of objects in the forest. However, to minimize replication traffic and keep the GC's database small, only selected attributes of each object are replicated. This is called the partial attribute set (PAS). The PAS can be modified by modifying the schema and marking attributes for replication to the GC.[12] Earlier versions of Windows used NetBIOS to communicate. Active Directory is fully integrated with DNS and requires TCP/IP—DNS. To be fully functional, the DNS server must support SRV resource records or service records.

Replication

Active Directory replication is 'pull' rather than 'push', meaning that replicas pull changes from the server where the change was effected.[13] The *Knowledge Consistency Checker* (KCC) creates a replication topology of *site links* using the defined *sites* to manage traffic. Intrasite replication is frequent and automatic as a result of change notification, which triggers peers to begin a pull replication cycle. Intersite replication intervals are typically less frequent and do not use change notification by default, although this is configurable and can be made identical to intrasite replication.

Each link can have a 'cost' (e.g., DS3, T1, ISDN etc.) and the KCC alters the site link topology accordingly. Replication may occur transitively through several site links on same-protocol *site link bridges*, if the cost is low, although KCC automatically costs a direct site-to-site link lower than transitive connections. Site-to-site replication can be configured to occur between a *bridgehead server* in each site, which then replicates the changes to other DCs within the site. Replication for Active Directory zones is automatically configured when DNS is activated in the domain based by site.

Replication of Active Directory uses Remote Procedure Calls (RPC) over IP (RPC/IP). Between Sites SMTP can be used for replication, but only for changes in the Schema, Configuration, or Partial Attribute Set (Global Catalog) NCs. SMTP cannot be used for replicating the default Domain partition.[14]

Database

The Active Directory database, the *directory store*, in Windows 2000 Server uses the JET Blue-based Extensible Storage Engine (ESE98) and is limited to 16 terabytes and 2 billion objects (but only 1 billion security principals) in each domain controller's database. Microsoft has created NTDS databases with more than 2 billion objects.[15] (NT4's Security Account Manager could support no more than 40,000 objects). Called NTDS.DIT, it has two main tables: the *data table* and the *link table*. In Windows Server 2003 a third main table was added for security descriptor single instancing.<ref name="blogs.technet.com"/

Programmatic interface

The features of Active Directory may be accessed programmatically via the COM interfaces provided by *Active Directory Service Interfaces*.[16]

Single server operations

Flexible Single Master Operations Roles (**FSMO**, sometimes pronounced "fizz-mo") operations are also known as operations master roles. Although domain controllers allow simultaneous updates in multiple places, certain operations are supported only on a single server. These operations are performed using the roles listed below:

Role Name	Scope	Description
Schema Master	1 per forest	Schema modifications
Domain Naming Master	1 per forest	Addition and removal of domains if present in root domain
PDC Emulator	1 per domain	Provides backwards compatibility for NT4 clients for PDC operations (like password changes). The PDC runs domain specific processes such as the Security Descriptor Propagator (SDPROP), and is the master time server within the domain. It also handles external trusts, the DFS consistency check, holds current passwords and manages all GPOs as default server.
RID Master	1 per domain	Allocates pools of unique identifiers to domain controllers for use when creating objects
Infrastructure Master	1 per domain/partition	Synchronizes cross-domain group membership changes. The infrastructure master cannot run on a global catalog server (GCS) (unless all DCs are also GCs, or environment consists of a single domain).

Trust

To allow users in one domain to access resources in another, Active Directory uses trusts.[17]

Trusts inside a forest are automatically created when domains are created. The forest sets the default boundaries of trust, and implicit, transitive trust is automatic for all domains within a forest.

Terminology

One-way trust

: One domain allows access to users on another domain, but the other domain does not allow access to users on the first domain.

Two-way trust

: Two domains allow access to users on both domains.

Trusting domain

: The domain that allows access to users from a trusted domain.

Trusted domain

: The domain that is trusted; whose users have access to the trusting domain.

Transitive trust

: A trust that can extend beyond two domains to other trusted domains in the forest.

Intransitive trust

: A one way trust that does not extend beyond two domains.

Explicit trust

: A trust that an admin creates. It is not transitive and is one way only.

Cross-link trust

An explicit trust between domains in different trees or in the same tree when a descendant/ancestor (child/parent) relationship does not exist between the two domains.

Shortcut

Joins two domains in different trees, transitive, one- or two-way.

Forest

Applies to the entire forest. Transitive, one- or two-way

Realm

Can be transitive or nontransitive, one- or two-way

External

Connect to other forests or non-AD domains. Nontransitive, one- or two-way.[18]

Windows Server 2003 introduced the *forest root trust*. This trust can be used to connect Windows Server 2003 forests if they are operating at the 2003 forest functional level. Authentication across this type of trust is Kerberos based (as opposed to NTLM). Forest trusts are transitive for all the domains in the trusted forests. Forest trusts, however, are not transitive.

Lightweight Directory Service

Active Directory Lightweight Directory Service (*AD LDS*), formerly known as *Active Directory Application Mode* (ADAM),[19] is a light-weight implementation of Active Directory. AD LDS is capable of running as a service on computers running Microsoft Windows Server. AD LDS shares the code base with Active Directory and provides the same functionality as Active Directory, including an identical API, but does not require the creation of domains or domain controllers.

Like Active Directory, AD LDS provides a *Data Store* for storage of directory data and a *Directory Service* with an LDAP *Directory Service Interface*. Unlike Active Directory, however, multiple AD LDS instances can be run on the same server.

Unix integration

Varying levels of interoperability with Active Directory can be achieved on most Unix-like operating systems through standards-compliant LDAP clients, but these systems usually do not interpret many attributes associated with Windows components, such as Group Policy and support for one-way trusts.

Third-parties offer Active Directory integration for Unix platforms (including UNIX, Linux, Mac OS X, and a number of Java- and UNIX-based applications), including:

- *Centrify DirectControl* (Centrify Corporation) – Active Directory-compatible centralized authentication and access control[20]
- *Centrify Express* (Centrify Corporation) – A suite of free Active Directory-compliant services for centralized authentication, monitoring, file-sharing and remote access
- *UNAB* (Computer Associates)
- *TrustBroker* (CyberSafe Limited) – An implementation of Kerberos
- *PowerBroker Identity Services*, formerly *Likewise* (BeyondTrust, formerly Likewise Software) – Allows a non-Windows client to join Active Directory[20]
- *Authentication Services* (Quest Software)
- *ADmitMac* (Thursby Software Systems)[20]
- *Samba* – Can act as a domain controller[21] [22]

The schema additions shipped with Windows Server 2003 R2 include attributes that map closely enough to RFC 2307 to be generally usable. The reference implementation of RFC 2307, nss_ldap and pam_ldap provided by PADL.com, support these attributes directly. The default schema for group membership complies with RFC 2307bis (proposed).[23] Windows Server 2003 R2 includes a Microsoft Management Console snap-in that creates and edits the attributes.

An alternate option is to use another directory service such as 389 Directory Server (formerly Fedora Directory Server, FDS), eB2Bcom ViewDS v7.1 XML Enabled Directory or Sun Microsystems Sun Java System Directory Server, which can perform two-way synchronization with AD and thus provide a "deflected" integration, as Unix and Linux clients authenticate to this while Windows Clients authenticate to AD. Another option is to use OpenLDAP with its *translucent* overlay, which can extend entries in any remote LDAP server with additional attributes stored in a local database. Clients pointed at the local database see entries containing both the remote and local attributes, while the remote database remains completely untouched.

Active Directory can be automated by Powershell.[24]

See also

- Active Directory Explorer
- AGDLP (implementing role based access controls using nested groups)
- Directory Services Restore Mode
- Flexible single master operation
- FreeIPA
- List of LDAP software

References

[1] "Active Directory on a Windows Server 2003 Network" (http://technet.microsoft.com/en-us/library/cc780036(WS.10).aspx#w2k3tr_ad_over_qbjd). *Active Directory Collection*. Microsoft. 13 March 2003. . Retrieved 25 December 2010.

[2] *Windows Server 2003: Active Directory Infrastructure*. Microsoft Press. 2003. pp. 1–8–1–9. ISBN 0-7356-1438-5.

[3] "Managing Sites" (http://technet.microsoft.com/en-us/library/bb727051.aspx). Microsoft CorporationA. . "An Active Directory site object represents a collection of Internet Protocol (IP) subnets, usually constituting a physical Local Area Network (LAN)."

[4] "Organizational Units". Microsoft Corporation. 2011. "An organizational unit in Active Directory is analogous to a directory in the file system"

[5] Microsoft Server 2008 Reference, discussing shadow groups used for fine-grained password policies: http://technet.microsoft.com/en-us/library/cc770394%28WS.10%29.aspx

[6] "Specifying Security and Administrative Boundaries" (http://technet.microsoft.com/en-us/library/cc755979(WS.10).aspx). Microsoft Corporation. 23 January 2005. . "However, service administrators have abilities that cross domain boundaries. For this reason, the forest is the ultimate security boundary, not the domain."

[7] "Sites overview" (http://technet.microsoft.com/en-us/library/cc782048(WS.10).aspx). Microsoft Corporation. 21 January 2005. . "A site is a set of well-connected subnets."

[8] "Planning for domain controllers and member servers" (http://technet.microsoft.com/en-us/library/cc737059(WS.10).aspx). Microsoft Corporation. 21 January 2005. . "[...] member servers, [...] belong to a domain but do not contain a copy of the Active Directory data."

[9] "Directory data store" (http://technet.microsoft.com/en-us/library/cc736627(WS.10).aspx). Microsoft Corporation. 21 January 2005. . "Active Directory uses four distinct directory partition types to store [...] data. Directory partitions contain domain, configuration, schema, and application data."

[10] Andreas Luther. "Active Directory Replication Traffic" (http://technet.microsoft.com/en-us/library/bb742457.aspx). Microsoft Corporation. . Retrieved 26 May 2010. "The Active Directory is made up of one or more naming contexts or partitions."

[11] "What Is the Global Catalog?" (http://technet.microsoft.com/en-us/library/cc728188(WS.10).aspx). Microsoft Corporation. 10 December 2009. . "[...] a domain controller can locate only the objects in its domain. [...] The global catalog provides the ability to locate objects from any domain [...]"

[12] "Attributes Included in the Global Catalog" (http://msdn.microsoft.com/en-us/library/ms675160(VS.85).aspx). Microsoft Corporation. 26 August 2010. . "The isMemberOfPartialAttributeSet attribute of an attributeSchema object is set to TRUE if the attribute is replicated to the global catalog. [...] When deciding whether or not to place an attribute in the global catalog remember that you are trading increased replication and increased disk storage on global catalog servers for, potentially, faster query performance."

[13] "What Is the Active Directory Replication Model?" (http://technet.microsoft.com/en-us/library/cc737314(WS.10).aspx). Microsoft Corporation. 28 March 2003. . "Domain controllers request (pull) changes rather than send (push) changes that might not be needed."

[14] "What Is Active Directory Replication Topology?" (http://technet.microsoft.com/en-us/library/cc775549(WS.10).aspx). Microsoft Corporation. 28 March 2003. . "SMTP can be used to transport nondomain replication [...]"

[15] efleis (8 June 2006). "Large AD database? Probably not this large" (http://blogs.technet.com/efleis/archive/2006/06/08/434255.aspx). Blogs.technet.com. . Retrieved 20 November 2011.

[16] Active Directory Service Interfaces (http://msdn.microsoft.com/en-us/library/aa772170(VS.85).aspx), Microsoft

[17] "Domain and Forest Trusts Technical Reference" (http://technet.microsoft.com/en-us/library/cc738955(WS.10).aspx). Microsoft Corporation. 28 March 2003. . "Trusts enable [...] authentication and [...] sharing resources across domains or forests"

[18] "How Domain and Forest Trusts Work" (http://technet.microsoft.com/en-us/library/cc773178(WS.10).aspx). Microsoft Corporation. 3 June 2006. . Retrieved 1 June 2010. "Defines several kinds of trusts. (automatic, shortcut, forest, realm, external)"

[19] "AD LDS" (http://msdn.microsoft.com/en-us/library/aa705886(VS.85).aspx). Microsoft. . Retrieved 28 April 2009.

[20] Edge, Charles S., Jr; Smith, Zack; Hunter, Beau (2009). "ch. 3". *Enterprise Mac Administrator's Guide*. New York City: Apress. ISBN 9781430224433.

[21] "Samba4/Releases/4.0.0alpha13" (http://wiki.samba.org/index.php/Samba4/Releases/4.0.0alpha13). *SambaPeople*. SAMBA Project. Archived (http://web.archive.org/web/20101115160233/http://wiki.samba.org/index.php/Samba4/Releases/4.0.0alpha13) from the original on 15 November 2010. . Retrieved 29 November 2010.

[22] "The great DRS success!" (http://people.samba.org/people/2009/10/05#drs-success). *SambaPeople*. SAMBA Project. 5 October 2009. Archived (http://web.archive.org/web/20091013094528/http://people.samba.org/people/2009/10/05) from the original on 13 October 2009. . Retrieved 2 November 2009.

[23] "RFC 2307bis" (http://www.padl.com/~lukeh/rfc2307bis.txt). . Retrieved 20 November 2011.

[24] "Active Directory Administration with Windows PowerShell" (http://technet.microsoft.com/en-us/library/dd378937(WS.10).aspx). Microsoft. . Retrieved 7 June 2011.

External links

- Microsoft's Active Directory Page (http://www.microsoft.com/windowsserver2008/en/us/active-directory.aspx)
- Microsoft Technet: White paper: Active Directory Architecture (http://technet.microsoft.com/en-us/library/bb727030.aspx) (Single technical document that gives an overview about Active Directory.)
- Microsoft Technet: Detailed description of Active Directory on Windows Server 2003 (http://technet.microsoft.com/en-us/library/cc782657(WS.10).aspx)
- Microsoft MSDN Library: [MS-ADTS]: Active Directory Technical Specification (http://msdn.microsoft.com/en-us/library/cc223122.aspx) (part of the Microsoft Open Specification Promise)
- Active Directory Application Mode (ADAM) (http://technet2.microsoft.com/windowsserver/en/library/6a4ae393-17e9-425d-b619-969dab507a6d1033.mspx?mfr=true)
- Powershell scripts to search Active Directory (http://www.happysysadm.com/2010/07/powershell-script-to-search-active.html)

Article Sources and Contributors

MassTransit_Enterprise *Source*: http://en.wikipedia.org/w/index.php?title=MassTransit_Enterprise *Contributors*: Alpha Quadrant, Bearcat, Chowbok, Sudhir h, TReidLewis, Woohookitty, 1 anonymous edits

Managed_file_transfer *Source*: http://en.wikipedia.org/w/index.php?title=Managed_file_transfer *Contributors*: AJPayne, Accellion, AngelSL, AnttiToivanen, Artemgy, BrownHairedGirl, DanielRigal, Dirkzwart, Dreznvc, EDIPM, Eeekster, Eks ssi, Elmorgan, Esalkin, Etamar, Fluxdavid, Icyganov, Intgr, IpswitchFT, Isarian, Jcbbl, Jschlauch, Kku, Managed Phil, Maurizio.simonetti, Mbucey, Mft2011, Mick09887, Mild Bill Hiccup, Mindmatrix, MrOllie, Phillidev, Radiantmatrix, Rich Farmbrough, Ronz, Rwwww, Securefiles, SharpeIT, Southrivertech, Sporcus8, Sprasad ganti, StephenBuxton, Susanpadura, TReidLewis, Terrible Tim, Tohd8BohaithuGh1, Tskybakmoen, Tyngman15, W Nowicki, Wareagles22, Wikidemon, 44 anonymous edits

GroupLogic *Source*: http://en.wikipedia.org/w/index.php?title=GroupLogic *Contributors*: Alpha Quadrant, D6, ErikTheBikeMan, Hmains, Iammike, J.delanoy, Jessemv, Kimsmith36, Peter E. James, Squids and Chips, TReidLewis, Waacstats, Woohookitty, 5 anonymous edits

Macintosh *Source*: http://en.wikipedia.org/w/index.php?title=Macintosh *Contributors*: 0pen$0urce, 1+2x3=9, 12.234.49.xxx, 209.75.42.xxx, 212.53.104.xxx, 21655, 24.15.135.xxx, 24.45.80.xxx, 2D, 32-bit dirty?, 3Coins, 4piecemcnugget, 63.192.137.xxx, 7, 75tpickupsx1983, 790314002, ?alabio, A bit iffy, A100nonymous, A2-computist, A8UDI, Aaron Schulz, Abce2, Abebenjoe, Abenjoel13, Abrech, Academic Challenger, Accurizer, Acps110, Acroterion, Adamdavid85, Adashiel, Adbmice, Addshore, Adjkabeef, AdjustShift, Adjusting, Admrboltz, Adz-834, Ae-a, Aerialvendetta, Aeryck89, Aetheling, Afknlnmfs, Ageekgal, Agendum, Ahoerstemeier, Aido2002, Airplaneman, Aitias, Akamad, Akcarver, Alainr345, Alansohn, AlbertR, Aleenf1, Alegoo92, Alex Heinz, Alex.muller, Alex22351, Alexf, Alexius08, AlexiusHoratius, Ali@gwc.org.uk, Alison, AlistairMcMillan, Allmightyduck, Allstarecho, Alpha Rex, Alphachimp, Altenmann, Althepal, Amatulic, Ameliorate!, Amyroe200, Anas 1762, AndersL, Andres, AndrewHowse, Andrewlp1991, Andrewpmk, Andy pyro, Andypandy.UK, Angela, Angelic Wraith, Angelsfreeek, Angr, AnonHat, Anonymous 198736, Anonymous 57, Anonymous 94056, Anonymous Paul, AnsonMBO, Anss123, Antandrus, Ante Aikio, Anthony, Antimac, Anville, Anvish, Apparition11, Apple2gs, Applehockey87, Applesmellz, Appleuser4, AquaStreak, Aquietlittlecat, Arbero, ArchStanton69, Arcyf, AriX, Arjunpachory, ArnoldReinhold, Arrista30, Arru, Ash, Aspects, Astral, Astronautics, Asyndeton, Athomsfere, Atlant, Aude, AugieD369, AussieBoy, Austin Hair, Avicennasis, Avoided, Avs5221, Awesome.js, AxelBoldt, Ay98182, Az1568, AzaToth, B, Backmaybe, Baconcheeseburger, Bad Bad Guy, Bagatelle, Bajajvikram, Barntks, Barrylb, Baryonic Being, Bassplaya7, Bateau, Bathrobe, Batman tas, Batmanand, Bazymac, Bbatsell, Bbpen, Bbriggs1, Bdude, Beao, Beastlyskillzz, Beau-lane, Beefman, Beeswaxcandle, Belovedfreak, Ben Ben, Ben Boldt, Ben-Zin, BenBurch, Benscripps, Bento00, BetaGamma, Betta Splendens, Bettia, Bevo, Bfigura's puppy, Bhadani, Bheron, Bhny, BigBrightStars, BigBurkey, BigDunc, BigFatBuddha, BigHaz, Biglovinb, Bill PLates, BillFlis, Blainster, BlaiseFEgan, Blanchette, Blood sliver, Bloodshedder, BlueBeach, Bluedude5887, Blueking12, Bluhd, Bmdavll, Bob Schaefer, Bobblewik, Bobdobbs1723, Bobo192, Bobthechick, Boing! said Zebedee, Boner2k8, Bongwarrior, BonzaiRob, Booieh, Bordgious, Boredzo, BorgQueen, Borisavljevic, Borisborf, Bpazolli, Bradv, Brian Kendig, Brian Patrie, Brian Tvedt, Brianga, Bricology, Brighterorange, Brion VIBBER, Brmwk, Brooklynl, Brossow, Bruce1ee, Brucethemoose, Brusselsshrek, Bryan Derksen, Bryanfosterthesexy, Bte288, Bubba73, Bubbadaboy, Bullzeye, BurtPeck, Burzmali, Bus stop, BusterD, BusterD public, Butterfly0fdoom, C S, C-squared, C.Fred, C628, CWY2190, CWii, Cab88, Cachdis123, Caltas, Calvin 1998, Cam493, CambridgeBayWeather, Camhusmj38, Camw, Can't sleep, clown will eat me, CanadianLinuxUser, Canderson7, CanisRufus, Canthusus, Captain Wikify, Carsten Levin, Carter4444, Caseybutt, Casmith 789, Catdude, Catgut, Catslash, Causa sui, Cayzle, Cbh, Cdespinosa, Cdmarcus, Cedars, Cei, Cenarium, CenozoicEra, Centurion123, Centurion31337, Chaffers, Chainsawriot, Chairboy, Chardish, Charles Edwin Shipp, Charles Gaudette, Charleschuck, Charlie Quebec Delta Echo Seven Sierra Foxtrot, Chaser, Checco, Chemmer, Chick Bowen, Chicocarlucci, Chillin69x, Chmod007, Choalbaton, Chris 73, Chrislk02, Chrisrowhatey, ChristTrekker, Christopher Parham, Chriswiki, Chubbles, CimanyD, Cinnamon42, Ciphergoth, Cjmarsicano, Ckatz, Clam0p, Classicfilms, Cleared as filed, Clevelandmusic24, Cliffb, Closedmouth, Cmdrjameson, Cnjartist, Cobbb, Coleflinn, Colin13, Collinrocks26, Cometstyles, Commit charge, CommonsDelinker, Compilation finished successfully, Computerdan000, Connectionfailure, Conner 36, Connormah, Conversion script, CoolFox, Coooltonnn., Cosmo0, Courcelles, Cpchase, Cpl Syx, Craft Bio, Crazysim, Creeper1342, Cregq, Crem23, Cremepuff222, Cretin's Bezoar, Crohnie, Cryosheath, Csaag, Csl77, Ctachme, Ctjf83, Cureden, Curps, CyberSkull, Cybercobra, Cyberia23, Cyktsui, DARTH SIDIOUS 2, DF08, DGJM, DHeyward, DMacks, DSGPS, DSLITEDS, Da Dude, DaL33T, DabMachine, Dale Arnett, Dalek king7, DalekClock, Damian Yerrick, Damicatz, Danallen46, Danamania, Danielfolsom, DannyDaWriter, Dannydream9999, Danrha, Danstein7, Dantheman531, Darcagn, Dark Shikari, DarkElrad, Darks0uldier, Darth Panda, Darth aavon, Darwinek, Dave Farquhar, Dave souza, David Fuchs, David Wahler, DavidSJ, Davidm617617, Dawnseeker2000, Dbfirs, DeadEyeArrow, Debatewise, Debresser, Deconstructhis, Deflective, Dekisugi, Delldot, Deltabeignet, Denimadept, Deor, DerHexer, Deus Ex, DevastatorIIC, Devon rodriguez, Dewet, Di Stroppo, Diagonalfish, Diemunkiesdie, Digana, Digi23azlan, Dims110, DirtPriest, Discospinster, Diyar se, Dlohcierekim, Dmlandfair, Dmsar, Dnstest, DocWatson42, Docski35, DoctorW, Dogguitar, Dogsandcatslivingtogether, Dominic Hardstaff, Dominic7848, Donreed, Doovinator, Doradus, Doulos Christos, Douthy, Download, Dr Debug, Dr Fell, Dragomiloff, Dream of Goats, DreamGuy, Dsyzdek, Dtunnicliff, Durin, Dustan1456, Dylan620, DylanW, Dysprosia, Dzubint, E. Sn0 =31337=, EGroup, ESkog, EVula, Ebbers.ja, Ebe123, Edgar181, Edgarde, Editoruk, Edward, Edwy, Ehurtley, Eighth90, Eiscir, El C, El Krem, ElAmericano, Elf, ElinorD, Elkman, Elliotjordan, Ellmist, Elmo456, Elvey, Encephalon, Enchanter, Encyclopedant, Energetic is francine@yahoo.com, Enric Naval, Eo01415, Epastore, Epbr123, Equendil, Eran of Arcadia, Ergbert, Ergosteur, Ergzay, Ericoides, Erik9, Escape Orbit, EscapingLife, Esong200, EthanolRules, Evercat, Everyking, Everything counts, Exia, Eyeareveeyeen, Falcorian, Fan-1967, Fang Aili, Favonian, Fede.Campana, Felimrules, Feneeth of Borg, Fenevad, FeralDruid, Fern2131, FernandoAires, Fg2, Fieldday-sunday, Figureskatingfan, Final88, Finlay McWalter, Fire 55, Fireborn, Fishtorte, Fiskars007, Fjord23, Flcelloguy, Flewis, Flowerpotman, Fluffernutter, Flydpnkrtn, Foant, For great justice., Fortunateson, Fourthords, Frazzydee, Frencheigh, Frequencydip, Freshacconci, Freyr, Frijole, Frungi, Fshepinc, Fun and games are fun, Funandtrvl, Funnybunny, Fuzy2K, Fvw, Fæ, G-O-D of W-A-R, GGreeneVa, GHe, GRAHAMUK, GSMR, Gadha, Gah4, Gaia Octavia Agrippa, Gaius Cornelius, Gakusha, Galor, Galoubet, Gareth Owen, Garion96, Garrett Albright, Garrett247, Garrettw87, Garrithklingel, Gary King, Gateman1997, Gateway122, Gaz9407, Gblaz, Gdo01, Generica, Geoff.green, Geologist001, Geometry guy, GeorgeLouis, GeorgeStepanek, Gexmofia, Ghatten, Gholam, Gil Dawson, Gilliam, Glacialfox, Glen, Glen Pepicelli, Glenn, GnuTurbo, Gogo Dodo, GoldRenet, Goldfish007, Golftheman, Goncalves229, GoogledIt?, Gracenotes, GraemeL, Graemesmith, Graham87, GregAsche, GregPanos, Griffin5, Grm wnr, GroundedZero, Gscshoyru, Gsg9378, Gunshoters, Gurch, Gurubanks, Gus, Guy Harris, Guy M, Guyinblack25, Gwernol, Gyounk, Gyre, H92, HELLO 101, HJ Mitchell, Hahahaidontgetit, Haham hanuka, Haikupoet, HamburgerRadio, HammerinHank, Hanselcat, HappyCricketer, Happydude22, Harley22xxx, HarryHenryGebel, Harryboyles, Haseo9999, Hateless, Hayabusa future, Hayne, Hdlafglh, Hdt83, Headbomb, Hellboy4311, Hen7713, HenryLarsen, Hephaestos, Heptapod, HereToHelp, Heron, Hetar, HiddenInPlainSight, Hiddenpwner, Hillbrand, Hippo X, Hiroshimaboomboomboom, Hoho832, Holodoctor1, Honytawksroprater, Hoolme, Hootentogger, Hopping cat, Hotlorp, Hrajt, HubmaN, Human.v2.0, HummadJ, Husond, HuwPrestatyn, Hyad, Hydnjo, Hydrargyrum, Hydrogen Iodide, IAMAHIPO ocolor, IG Slayer, Iain.dalton, Iamalexjones77, Iameukarya, Iamunknown, Ian The Great, Ian13, Ianblair23, Ianweller, Ickyelf, Ignatzmice, Igoldste, Ijbond, IlliterateSage, Ilse@, Iluvcapra, Ilyaroz, Imperial92, Impfac, IndulgentReader, Ineffable3000, Infodriveway, Inks.LWC, Insolectual, Intelligentsium, Intersting, Ioeth, Iohannes Animosus, Iridescent, Irishguy, IronGargoyle, Irvela, IsUsername, IstvanWolf, Itfcsam, Ivymike21, Ixfd64, J.delanoy, JCarriker, JDP90, JForget, JNW, JP Godfrey, JR98664, JThinger, JaGa, Jaavaaguru, Jabencarsey, Jacek Kendysz, Jack The Hat, Jackalope darko, Jackfork, Jackmorino, Jackol, Jacobolus, JagSeal, Jakekito, Jamcamuk, Jamiemcc, Janizary, Janke, Janus Shadowsong, Jared, Jason C.K., Jasonq, Jaxl, Jcw69, Jdude204, Jeff G., Jeffrey Smith, Jeremiahhsu, JeremyA, Jerrie32, JhastyII, Jhertel, Jim Douglas, Jim Shaud, JimEOme, Jimthing, Jleedev, JoanneB, JoaoRicardo, JodyB, JoeCool59, Joelbelknap1, Joemexico16, Joemexico17, JoeyJoeyJoey, John Price, John Quincy Adding Machine, JohnInDC, JohnRDaily, JohnSawyer, Johnleemk, Johnmarkos, Johnrpenner, Jojhutton, Jokestress, Jonnyboy911, Jonpaulusa, Joost, Jordandanford, Jordanduhawesomedude, Josh the Nerd, Josh59x, JoshEdgar, Jossi, Jothesmo, Jotting9566, Jouer au Fat, Jovianeye, Jpk, Jsephton, Jtalledo, Jtbean, Jtkiefer, Jtn1209, Jtuttle, Julesd, Jusdafax, Jutiphan, Jw21, Jwissick, Jyoyopaek, Jéské Couriano, KGasso, KJS77, Kakaman, Kakurady, Kalathalan, Kaly99, Kanags, Karenjc, Karmann, Kasskid, Katefan0, Kbolino, Kchishol1970, Keegan, Keilana, KelvSYC, KenshinWithNoise, Keremtezcan, Kesac, Kesh, Kevin chen2003, Keyser Söze, Kickapples, Kierenfahey, Kiko4564, Killthemac, King of Hearts, Kingpin13, Kiscica, Kissedbythesun, Kiteinthewind, Kjetil r, Klausness, Klilidiplomus, Knowhow, KnowledgeOfSelf, Knoxus, Koolkid2168, Korg, Kozuch, Kraetos, Krakeder9, Kross, Kubigula, KudulO, Kukini, Kungfuadam, Kuru, Kwamikagami, Kyoko, Kyphe, Kyro, L Kensington, L3vyk3v4nd4ll, LAX, LBHS Cheerleader, La goutte de pluie, Laaaaabob, Lacrimosus, Lame Name, Lars T., Law, Ldo, LeaveSleaves, Leer25, Leftus, LegitimateAndEvenCompelling, LeilaniLad, Leon7, Leszek Jańczuk, Lethe, Lexein, Lfaporsche2424, Libstooge 01, Liftarn, Lightdarkness, Lightlowemon, Lightmouse, Lijorijo, Lilj242, Limelightist, LinaMishima, LindsayH, Linnell, LittleOldMe, Liubei, Llhm, Localzuk, Logan598, Loganm10, Lolcoaster, Lord Vader, LordBleen, Lostchicken, Lotta12, LovesMacs, Lowenddan, Lproven, Lradrama, LtPowers, Luigiwax, Luk, Lukobe, Luna Santin, Lupin, Luumu, M.nelson, ME4236, MER-C, MFNickster, MKoltnow, MONGO, MORRISPC10, MPD01605, MSTCrow, Ma8thew, Mabdul, Mac sux, Mac128, Mac2333, Macaddct1984, Maccomputerssuck, Macinjosh87, Macintosh User, Macintoshrocks9, Macrhino, Maebmij, MaganT2k13, Magicaldave, Magister Mathematicae, Mahanga, Mailseth, Makaristos, Makeemlighter, Makesdark, Malimbar04, Malo, ManIsWhiteMonkeyIsBrown, ManfrenjenStJohn, Manoflast3, Manufracture, Mareino, Marek69, Mario91234, Marj Tiefert, Mark5677, MarkGallagher, Markonen, Marskell, Martin451, Martinman11, MastCell, Master Freeman, Mathfan, Mathonius, Matt Britt, Matt the mexo, Matthew2414, Matthewwilliamproctor, Mattisse, Mav, Max Naylor, Maxim, Maxis ftw, Mboverload, McDonaldsGuy, McDoobAU93, McSly, Mcclurd, Mdd4696, Mdwh, Melmann, Mentifisto, Mephistophelian, Mera james, Merlion444, Mhkmack, Michael Hardy, Michael K. Edwards, Michael Snow, Michael93555, Michaelas10, Michaelbusch, Michelle eris, MidgleyDJ, MidiMacMan, Mihai1025, Mika1h, Mike Rosoft, Mike hayes, Mike1, Mikeo, Miles, Millifolium, Mimigu, Minesweeper, Mintchocicecream, Mipadi, Miquonranger03, Mirek2, Mirror Vax, Miss Morning, Missmarple, Misza13, Mkay222, Mklbtz, Moe Epsilon, Moncrief, Mono, Moondyne, Moonriddengirl, MorningView375, Morten, Mproud, MrC539, MrClean03, MrKalamazoo2, MrRedwood, Mrclark411, Mrholybrain, Ms ArtGeek, Msikma, Mtk402, Muchclag, Mushroom, Muzzy136, Mvelez0002, Mvjs, Mwk teck, Myanw, Mysterioususer, Mystiq0, Mystyc, Mütze, N3rf360, NHRHS2010, Nakon, Nanshu, NapoliRoma, Nappilainen, NathanHurst, Nauticashades, NawlinWiki, Ndenison, Ne1, Neil D, Neilfein, Neoleeyennek, Neonleon123456789, Nerdd, NerdyScienceDude, Nesnej15, Netizen, Netminder, Netoholic, Neverquick, NewEnglandYankee, Newkai, Nice Dad, Nice yum felicia@yahoo.com, Nickgill1, Nicoanison, Niemasd, Nigholith, Night of the Big Wind, Nightscream, Nikbro, Nikorasu, Nintendros, Nishkid64, Niteman555, Nixdorf, Nlu, Noetica, Nolat, NorseLord, NosraePekuL, NotSuspicious, Nothlit, Nricardo, Nsaa, Nsayer, NuclearWarfare, Nuggetboy, Numberp, Nwwaew, ONUnicorn, Oberonfoxie, Obli, Oblivious, Oda Mari, Ohconfucius, Ohnoitsjamie, Ohwhatcoolk, Okma, Oldschool, Oli-Was-Here, Olivier, Omar abdallah, Omegatron, Omicronpersei8, Omondra, One, Oneupthextraman, Onorem, Onthost, Onza110, Optimist on the run, Optowiki, OrangeCrusader, OrbitOne, Orbitalwow, Oregonerik, Ori, OriginalGamer, Osxshortcuts, OverlordQ, Owlmonkey, Oxymoron83, Oyvind, Ozguy89, PC destroyer, PJM, PR3V3NTION, PS3 FTW, Pablomartinez, PageVandal, PanamaDrummer911, Pantergraph, Papadopa, Paraded junk soul, Parklandstudent, Particleman24, Patrick T. Wynne, Patrick0Moran, Patton76, Paul Magnussen, Paul3144, PaulGS, Paulkman, Paulmrussell12, Paxsimius, Payco, Pbroks13, Pdk, Peama058, Pedro, Peeka, Pekaje, Penfold1000, Permafrost, Persian Poet Gal, Petesmiles, Petri Krohn, Pgk, Phaedriel, Phantomboy98043, Pharaoh of the Wizards, Pharos, Phdrummer, PhilZ, Philipdw, Philiscool1919, Philosophicles, Philwelch, PhoenixVTam, Phuk you1, Physman, Pidgeot, Piledhigheranddeeper, Pill, Pilotguy, Pineapple1, Pinkadelica, Pjoef, Planetary Chaos, Pmurph, Po panda, PoccilScript, Pol098, Polarbowler, Polarit, PoliticalJunkie, Poopshnicles, Poplggp, Porqin, Possum, Praetorian42, Prari, Primate, Prodego, ProhibitOnions, Prolog, Proud,

Psantora, Pschemp, PseudoSudo, Psychoman247, Ptaylor270782, Puji6, Purinflash, Pyryle, Python eggs, Pyxzer, Qinnyc, QofASpiewak, Quadell, Quantpole, QuiteUnusual, Quuxplusone, Qwe, Qwerty1211, Qwertyiscool1234567, Qxz, R'n'B, R.carroll, RAMChYLD, RCpatroler, RadicalBender, Raffaele Megabyte, Raidon Kane, RainbowOfLight, Ramallite, Rambo23, Ramunnoodle, Ramunnoodles, Ramurf, RandomStringOfCharacters, RandomUsefulKnowledge, Randum, Rattlesnake, RattusMaximus, Raul654, RaviC, Rbj, Rcandelori, Rchamberlain, Rdikeman, Rdsmith4, Reaper Eternal, Rebroad, Red, RedHillian, RedSpruce, RedWolf, Reddi, Redeming, Redgreen88, Reedy, Registeraccount, Reisio, Remember the dot, Remstar, Repetition, Res2216firestar, Retsoperedevlebrm, Revolution221, RexNL, Rich Farmbrough, Richard D. LeCour, Richardcavell, Rigel1, Rita Moritan, Riverwaste, Rjairam, Rjwilmsi, Rklawton, Rnbangras, Robb0082, Robert Brockway, Robert P. O'Shea, RobertG, RobertMfromLI, Rodhullandemu, Rogerd, Romansanders, Ronhjones, Ronk01, Room01, Rory096, Roryethanr, RossPatterson, Rossami, Rostz, RoyBoy, Rpresser, Rrburke, Rschen7754, Rsholmes, Rugby471, Rx 100e, RxS, Ryan Norton, RyanCross, RyanGFilm, Ryanaxp, Ryguy999, Ryrivard, S, SCJohnson77, SD6-Agent, SMC, ST47, Saalstin, Safarirss6394, Salanus, SallyForth123, Salty!, Sam Korn, Sandridge, SandyGeorgia, Sango123, Saptahrshi, Sardur, Sarranduin, Save-Me-Oprah, Scanlan, Scepia, Sceptre, SchfiftyThree, Schissel, Schmidtmandaddy, SchnitzelMannGreek, School99144, Schumi555, SchuminWeb, Schutz, Scientus, Sciurinæ, ScottSteiner, ScottlovesFanta, Scrotie, Scrumshus, Scwlong, Scythre, Sdedeo, SeaFox, Sebapeluca, Seidenstud, Seinfreak37, SenorKristobbal, Seth ze, Severusskad, Shadowlynk, ShaneKing, Shanel, Shaneo95, Shanes, Shanesan, Shannon Bradley, Shenme, Shinjiman, Shirik, Shizane, Shoeofdeath, Shoy, Simeobindo, Simmonsjd02, Simpsonfan25, Singapore Jones, Sionus, Sir Anon, Sir Nicholas de Mimsy-Porpington, Sir Stig, SirJibby, Skier Dude, Skunkboy74, Skyezx, Slady, Slakr, Slark, Sleigh, Slugger, Smalljim, Smyth, Snigbrook, Snowolf, SoCalSuperEagle, Sobrow, SolFruit, Somebody in the WWW, Someguy1221, Something12356, Sophisticated 23, Sora089, Soumyasch, Sp, SpaceFlight89, SpaceRocket, Spacesoftware, Spartan-James, Spencerperry, SpigotMap, Splash, SpookyMulder, Spritzsinger, SpuriousQ, Spyzy, Srikeit, Stan Shebs, Staney23, Starnestommy, Statsong, Steekel, Stefan, Stephan Leeds, Stephen Shaw, Stephenb, SteveRamone, SteveSims, Stevefis, Stevejobsforpresident, Steven Zhang, Stevenj, Stevenrasnick, StrawBerryBannaCream, Sturm55, Subsailor, Suffusion of Yellow, Suhailpeerbhai, Summitstudent, SuperHamster, Superdude876, Surv1v4l1st, Sverdrup, Switchfoot, Symode09, SyntaxError55, Syrthiss, SystemBuilder, TBSchemer, TDS, TEG24601, THEN WHO WAS PHONE?, TWCarlson, Takuzinis, Tangentidea, Tango, Tangotango, Taroaldo, Tarquin, Tassedethe, Tau'olunga, Tawker, Taxman, Tbhotch, Tbutz, Tcncv, Tdog106106, Teacish, Tech30, Techpodusa, Teles, Tempshill, TenPoundHammer, Terrillja, TexasAndroid, Tgeairn, Thakur rigved, The Catfish, The Cunctator, The Gilly, The Grate Spanish Duck, The Interior, The Land of Smeg, The Mysterious El Willstro, The Parting Glass, The Rambling Man, The Thing That Should Not Be, TheDJ, TheDoctorZoidberg, TheSheriff2004, Theanthrope, Theblueslime, Thechoirlife2006, Thehelpfulone, Thejadefalcon, Thejaggedtimes, Thesab, Thespencerjones, Thewallowmaker, Thg250, Thingg, Thomas Ash, Thomas Gilling, ThomasHarte, ThomasK, Thortful, Thursdayx20, Tide rolls, TigerK 69, TigerShark, Tim1357, Timrollpickering, Titoxd, Tktktk, Tman8809, Tmarki, Tobacco bean, Toddan, Tom Morris, Tommy2010, Tony1, TonyClarke, Tooki, Toonmon2005, Tooza64, Torbix, Tosha, TotalSpaceshipGuy3, Totie, Toussaint, Toytoy, Tpbradbury, Tragen1234, Trailbum, Travelbird, Treaeakjdsfaew4h'iiioii, Tree Falling In The Forest, Trevor MacInnis, TrevorLSciAct, Trickybiz, Triwbe, Troy 07, Truenaruto, Trusilver, Turkeybacon97, Tutnkmn, Tuxisuau, Tverbeek, Twang, TwilligToves, Tyler, Typhoon, Tyw7, Ubcule, UberPwnage92, UchihaKazu, Ufosjt, Ulric1313, Ultra Galahad, UncleBubba, UncleDouggie, Unreal7, Unyoyega, Upi, Usb10, Usgnus, Uzsa, V-train, VMS Mosaic, Valermos, Validation7, Vanderbeekfan13917, Vary, VegKilla, VegaDark, Veinor, Veldin963, Velella, Venom098, Versus22, VibhuC, Vikramc84, Vizcarra, Vranak, Vrenator, Vulpes vulpes, WJSProwler, Wackymacs, Wakleon, Wampa Jabba, Wapcaplet, Warren, Wartrakk, Wavelength, Waycool27, Wayward, Wcquidditch, Webeditor909, Wechselstrom, Wendell, Wereon, Wernher, Wertjoe, West Brom 4ever, West.andrew.g, Wgungfu, Whale, Whitebox, Whywhenwhohow, Wifione, Wiki Roxor, Wiki alf, Wiki b00b12, Wiki man120, WikiED, WikiLaurent, WikiLegend12, Wikimassiker, Wikipedian456, Wiknerd, WillMcC, William Avery, Wimt, Win2000Pro, WindowzRULZlolZ, Wine Guy, Winston007, Wintersun777, Wipsenade, Wjmummert, Wknight94, Woodsstock, Woodwynlane, Woofs, Wpedzich, Wrcmills, Wtmitchell, Ww2censor, X!, X570, Xaosflux, Xdriver206, Xelentoliver, Xeworlebi, Xiggle, XuperUber, Yakbasser, Yaksar, Yamen993, Yamla, Yankeezfan1, Yeagh, Yiwenhuang113, Yoasif, Youatemypotatoes, Yuhong, Zacherik, Zackkattackk247, Zackyboy96, Zalinda Zenobia, Zarmanto, Zebruh, Zekeman95, Zig973, Zinoviev, Zisa, Zlemming, Zm1573, Zoicon5, Zolotros, Zomganairship, Zondor, ZooFari, Zschernicky, Zsinj, Zwynky, Zzuuzz, Ævar Arnfjörð Bjarmason, Œ∑´®†¥πø^, مل اع بوبحم, 2849 anonymous edits

Microsoft_Windows *Source*: http://en.wikipedia.org/w/index.php?title=Microsoft_Windows *Contributors*: -Majestic-, 03jmgibbens, 1(), 16@r, 1nt2, 2mcm, 2toy mora, 62.253.64.xxx, 67773732TYU, 68DANNY2, 9ms, A Raider Like Indiana, A gnome, A-giau, A7x, ACCOM2222, Aanhorn, Abhishikt, AbsoluteFlatness, Academic Challenger, Acroterion, Adam Mirowski, Adamacious, Adamodell, Admin@pcrevs.com, Ae-a, Aeæ, Afro Article, Ageha Winds, Ahodacsek, Ahoerstemeier, Aidan W, Aido2002, Aihtdikh, AimalCool, Airplaneman, Ajm81, Akamad, Akhristov, Akira-otomo, Aksi great, Albert0057, Aldie, Alegoo92, Alemily, Alfio, AlistairMcMillan, Allstarecho, Almafeta, Alpha 4615, Alphathon, Alphius, Alshaheen15, Althepal, Am088, Amakuru, Amcfreely, Amrykid, Andre Engels, Andrevan, Andrewtechhelp, Andros 1337, Andy16666, Andyh2, Angeljon121, Ann Stouter, Anog, Anonymous Dissident, Anonymous56789, Antandrus, Anthall1991, Antimatter15, Antique Rose, Anville, Anþony, Aomarks, Aqair, Aranel, Arch dude, Archer3, Archivist, Aria1561, Arnoldkul, Arthena, Ashdurbat, AstroNomer, Astroview120mm, Aude, Audrius u, AussieLegend, AvantgardeMVC, Aveilleux, Avenue, Avihut, Avocado27, Ayjay1545, Azrael Nightwalker, B, BNSF Man, BUzTeD, Badwolf2212, Bakery2k, Banes, Barek, Barrera marquez, Bartosz, Bbq man, Bbriggs1, Bdoserror, Bdshort1, Beao, Beinsane, Ben-Zin, BenBen1234, Benandorsqueaks, Benc, Berek, Bergsten, Betelgeuse, Bevo, Bhadani, Bibliomaniac15, Big Bird, Big Booger, BigCow, BillG, BillWSmithJr, Billyswong, Bimmerosx, Binsurf, Binzisimpsons, Bissinger, Blackanddarkness, Blackcap95, Blackcats, Blakkandekka, Blaxthos, Bleakcomb, Blobglob, Blowdart, Blubberboy92, BlueCaper, Blueforce4116, Bluestriker, Bo98, Bobo192, Bobwrit, Bogods, Bollyjeff, Boris Allen, BostonMA, Boylett, Brandizzi, Brandon Brown, Brian0918, BrianGo28, BrianRecchia, Brianski, Brisvegas, BrokenSegue, Bubba hotep, Buchanan-Hermit, Buddha24, Bunnyhop11, Burntsauce, CRFWNY, CWenger, Cacophony, Cadiomals, Caffelice, Caltas, Camembert, Can't sleep, clown will eat me, Canadian-Bacon, Candamir, Candorwien, CanisRufus, Caper13, CaptainVindaloo, Casper10, Casper2k3, Cayindra, Cbrown1023, Cburnett, Ceeon, Celestra, CesarB, Ceyockey, Cff12345, Cgnabod, Cgs, Charles Gaudette, Charles dye, Chazz, Cherkash, Chille, Chmpoure, Chocolatemilk94, Choppie3000, Chowbok, Chris 73, Chris Pickett, Chris the speller, Christian List, Chriswoz, Ciao 90, CitronManden, Cityscape4, Cjcamilla, Cjordan93, ClamDip, ClockworkTroll, Clyde1998, Cncccer, Cncxbox, Codificate, Codyblevins, Coffee, Colejohnson66, Colin Hill, Cometstyles, Commander Keane, Computer boy2, Computerdan000, ConCompS, Coniosis, Conman23456, Conversion script, Coolcaesar, CoolingGibbon, Corporal clegg48, Courcelles, Cp111, Cpiral, Crazycomputers, Crazyromo, Crem23, Cremepuff222, Cristan, Crpietschmann, Cryptic, Cstanners, Ctbolt, Cuvtixo, Cvinoth, Cwolfsheep, Cyktsui, Cynical, Cyrius, D, D thadd, DHN, DOHC Holiday, DaDrumBum, DalekClock, Damian Yerrick, Dan D. Ric, Daniel5127, DanielRigal, Danny Beaudoin, DarkFalls, DarkHorizon, Darranc, Darremon, Darthnader37, Dasani, DataMatrix, Databases, Davelane, David Biddulph, David Gerard, David1409, Davidjk, Dcandeto, Dcoivin, Dcshank, Deanhowell123, Debackerl, Deeahbz, Dehumanizer, Delirium, Delldot, Deltabeignet, Demmy, Denniss, Derek Parnell, Deryck Chan, Dethomas, Deus2, Dieboybun, Diegogrez, DigbyDalton, Digita, DigitalLife, Dimre01, Dina, Dinhtuydzao, Dinjired, Dinnerface, DinosaursLoveExistence, Disorganisation Man, Dj789, Djegan, Djhybrid117, Djmasala, Dmerrill, DmitryKo, Dog1818, Dojarca, Dolphinn, Downloaddude1258, Dp462090, Dragar Gt, Dragon 280, DragonflySixtyseven, Dragonhelmuk, Dstln, Dtech, Dudesleeper, Dungodung, Dust Filter, Dustin gayler, Dwheeler, Dylan Flaherty, Dysprosia, Dzubint, EH74DK, Eagleal, Ed g2s, Edgar181, Edgarde, EdgeOfEpsilon, Edward, Eclamstylez77, Eggsacute, Egil, Eivindsol, El Dominio, Elassint, Eliotwiki, ElliotAdderton, Ellmist, Elm-39, Eloquence, Elwood00, Emacsuser, Enco1984, Endofskull, Enigmaaaaa, Enno, Enochlau, Ente75, Eraserhead1, Ergosteur, Esanchez7587, Escape Orbit, Esebi95, Evice, Evil Monkey, Evil saltine, Evosoho, Ewlyahoocom, Exodite, F80, FF2010, Fabsss, Faisal.akeel, Faithlessthewonderboy, Falco McCloud, Fantasy, Faradayplank, FatalError, Fatjoe151, Favonian, FayssalF, Fbv65edel, Fdp, Felipe Aira, Fennec, Ferkelparade, Ffx, Fiskegalen92, Fitzebwoy, Fiver2552, Flamurai, Flanakin, FleetCommand, Flying Bishop, FlyingPenguins, Foo1995, Fowl2, FrancoGG, Frap, Fraslet, Freakofnurture, Frecklefoot, Fred Bradstadt, Frederik.Questier, FreeKresge, Freeeekyyy, Frenchman113, FreplySpang, FrozenPurpleCube, Frozenport, Fuck You, Funandtrvl, Furrykef, Fuzheado, GDonato, GNUtoo, GSK, Gabrielpokemon, Gadfium, Gaius Cornelius, Galactor213, Galoubet, Galwhaa, GamerXp, Gamerzworld, Gaming&Computing, Ganeshotaku, Ganfon, Gang65, Gaodifan, Gardar Rurak, Gary King, Gary Kirk, Gavin.perch, Gazpacho, Ged UK, Geek45, Geoffspear, George Adam Horváth, Georgia guy, Geraki, Gerbrant, Ghettoblaster, Gilbertogm, Gilliam, Glen, Gnepets, Goatasaur, Gobonobo, Gogo Dodo, Goldom, Googler459, Googlesucks30789, Gosub, Graham87, GrandPoohBah, Graysni, Gregory Heffley, Grin, Grm wnr, Groggy Dice, Gronky, Grunt, Gryllida, Gscshoyru, Gtdp, Gunnar Guðvarðarson, Gustyfalcon, Guy Harris, Guyjohnston, HJ Mitchell, Haakon, Haeleth, Hahafatpeople, Hairchrm, Hall Monitor, Halsteadk, Hammersoft, Harryboyles, Haseo9999, Hbomb phd mom, Hdt83, Heapchk, Hebrides, Helixblue, Hello32020, Hendrixski, Henry W. Schmitt, HereToHelp, Hirosho, Holbred, Howardjp, Hungupbg, Hurtstotalktoyou, Husky, Hydr, Hypnoticcyst, I have wood, I8189720, IGod, ILOVELOL32, IMSoP, IXella007, Iain99, Ialsoagree, Ian Pitchford, Ianjones50, Ichangefacts, Iggymwangi, Ilaiho, Illegal Operation, Illyria05, Ilya, IlyaHaykinson, Ilyanep, Indefatigable, Indon, Injust, Interframe, Ioprwe, Ipodsocool, Iridescent, IsUsername, ItsProgrammable, Ivan Pozdeev, Ixfd64, J Di, J36miles, JCarriker, JIP, JLaTondre, JWSchmidt, JYOuyang, JYolkowski, Jacob Hnri 5, JacquesStrap, Jaericho, Jake Wasdin, Jamieostrich, Jan Hofmann, Jasper Deng, Jauerback, Jaxl, Jcbparry, Jcurtin, Jdlowery, Jdm64, Jeffwang, Jeh, Jennie--x, Jeremy Visser, JeremyA, Jerryseinfeld, Jerrysmp, Jesse Viviano, Jessemv, Jesus5555, Jesus764, Jesuss, JettaMann, Jevel66, Jiffy, Jigs41793, Jimmi Hugh, Jimothytrotter, Jimthing, Jiy, Jketola, Jkonline, Jmath666, Jmjglick, Jmoynihan08hm77, JoJan, JoThousand, Joanjoc, JoanneB, Joe2832, Joejoejo, Joffeloff, Johann Wolfgang, John, John Ericson, John Goettle, John Quincy Adding Machine, JohnJamesWIlson, JohnOwens, Johnclow13, Johnleemk, Johnmc, Johnuniq, Jojit fb, Jon vs, JonasL, Jonatan Swift, Jondel, Jonghyunchung, Jopxton, Jordan015, Jordanyoung17, Jose Concepcion, Josh the Nerd, Joshua Issac, JoshuaArgent, Joyous!, Jpoke89, Jrcure, Jsmethers, Jtc, Ju66l3r, JuJube, JuanC08, Julian Mendez, Jumbo Snails, Justanother, Jyoz, Kakomu, Kar.ma, Karimarie, Karlo2002, Karnesky, Katarighe, Kaycubs, Kaysov, Kbh3rd, Kbolino, Kelly Martin, Kenny Strawn, Kesla, Kevinkor2, Keyser Söze, Khym Chanur, Kierenj, Kigali1, KingpinE7, Kinneyboy90, Kipholbeck, Kissmeplease, Kittyhawk2, Kjhonsa, Kkm010, Klingoncowboy4, KnowledgeOfSelf, Knutux, Kohlmalo, Kolonuk, Koman90, Kop, Korossyl, Kozuch, KrakatoaKatie, Krawi, KrisBogdanov, KumardipSarkar, Kungfuadam, Kungming2, Kurieeto, Kushared, Kwen, Kypr8, L Kensington, La goutte de pluie, Lacrimosus, Lahiru k, Lantay77, Larry laptop, Lbs6380, Lcarsdata, Lead$peaker, Leandrod, Lectonar, LeeHunter, Leif, Leithp, LeoNomis, Letdorf, Leuko, Liberty Miller, Lightmouse, Linear88, Linuxbeak, Linuxerist, LionKimbro, Llalala, Localzuk, Loismum, Lonaowna, LonelyBeacon, Looksliketrent, Lord British, Lordalpha1, Lotje, Lou.weird, Lowellian, Luce007, Luk, Lumpbucket, Luna Santin, Lupin, M Johnson, MADCastro2012, MCC, MER-C, MERC, MJ94, MJGR, MK8, MMMEEE, MZMcBride, Ma8thew, Mac, MacGeekGuy, Macurry, Madcow 93, Maester mensch, Magister Mathematicae, Mailer diablo, MainFrame, Malo, Man123123, Marceki111, Marcok, Mark, Mark85296341, MarkGallagher, MarkSpearmint, Markhurd, Markie2, Markustwofour, Martarius, Martyx, Marudubshinki, Marx Gomes, Marysunshine, Massysett, MasterCole, Mastershake86, Mathwizxp, Matt Crypto, Matt.forestpath, MattGiuca, Mattarata, Mav, Max Schwarz, Maxim Masiutin, Maximus Rex, McLovin34, Mcwatson, Meelar, Mehran, Mephiles602, Metalim, Methanegas, Metz2000, Mguy77, Mgw854, Michael Angelkovich, Michael Ray, Michaeldadmum, Michaelpremsrirat, MichealH, Midgrid, Midkay, Midnight Green, Mike4ty4, MikeZuniga, Mild Bill Hiccup, Milfman, Minesweeper, MinorItem, Miros 0571, Mithent, Mjpieters, Mkdw, Mlscdi, Mmoll21m, Mmoller, Mms, Mo0, Modster, Mokk123, Monedula, Monothleft, MontyB, Moonknightus, Moonridersaregay, Moonwolf14, Mr Bartels, Mr. Lefty, Mr. XYZ, Mrdelayer, Mrleewilliams, Ms2ger, Mugunth Kumar, Mulad, Mushroom, Mvmarier, Mwtoews, Mxn, Myanw, Nachmore, Naddy, Naltrexone, Nanshu, NapoliRoma, Nathan nfm, Nathanlilienthal13, Neckername, Neelchauhan, Neo The User, Neo2256, NewGuy4, Newfraferz87, Nicholas Love, Nick2253, Nickenzi, Nimc, Nintendog master 54, Nishkid64, Nixdorf, Njan, Nmnmnm, Noypi380, Nukleartoaster, Nurg, O, Odd bloke, Off!, Okwestern, OlEnglish, Olioster, Ollie the Magic Skater, Omicronpersei8, Ondrejsv, Oneupthextraman, Opagecrtr, Opelio, OregonD00d, Oscarthecat, Ost316, OtherPerson, Otter Mii-kun, OwenX, Oxymoron83, Ozherb, PPGMD, Packard Bell Legend, Padsquad43, Pakaran, Patrick, Patrickweeks007, Patrolman89, Paul Stansifer, Paulfp, Pavel Vozenilek, Pcuser42, Persian Poet Gal, Person1936, Petaluma Paranormal, Peter Grey, Peteturtle, Petrwiki, Peyman4u, Peyre, Pgiii, Pgk, Phanton, Phil Sandifer, Philip Trueman, Phoogenb, Pigman, Pile0nades, Pinikas, Piratesmack, Plugwash, PoWwOw123, Pokyrek, Poor Yorick, Poweroid, Prjrv787, Prolog, Propound, Pschulz01, PsyMar, Puffin, Puramyun31, Pursey, QUAZWRATH, Qantasplanes, Qudder, Quinsareth, Quoladdie, Qwerty124gg, Qwitchibo, Qxl32, Qxz, R'n'B, RB972, RTC, RadioActive, RadioFan, Radon210, Raja99, Ramanpotential, RandomP, Rasmasyean, Rbuj, Rchamberlain,

Rcmouse1010, Rd syringe, Rderijcke, Rdsmith4, Reallikeunreal, Rebroad, Red Dalek, Red Director, RedKlonoa, RedXII, Redvers, Reilly, Reisio, Remember the dot, Repetition, Resplendent, RevolverOcelotX, Rgbc2000, Rhesusmonkeyboy, Rhobite, Rhtc, Rich Farmbrough, Richard Lotspard, Richard626, Ricjl, RickK, Riluve, Rincewind32, Rjecina, Rjwilmsi, Rl, Robartin, Robbie940, Robert Buzink, Robert H, Robert Xia, RobertG, Robertd, Robinhw, Rocastelo, Rocket71048576, Rockymountains, Rodeo90, Roketjack, RolandH, Ronark, Rsantmann, Rsm99833, Running, Rursus, Rutherfordjigsaw, Ruud Koot, Rwwww, Ryan Norton, Ryan Postlethwaite, Ryan t moua, S0aasdf2sf, SF007, SHARD, SHeumann, SNIyer12, ST47, SYSS Mouse, Sae1962, Sam Hocevar, Samuel, Samuel Blanning, Samvscat, Sango123, Sasquatch, Sauronjim, Scepia, SchmuckyTheCat, SchuminWeb, Science4sail, Sciencegeek10188, Scientus, Scifiintel, Sciurinæ, Scorpiona, Scottymoze, Scriberius, Scuiqui fox, SeanMack, Sebmathews, Sebrat, Secretlondon, Selfdiscipline, Selivanow, September 11 terrorist, Sethoeph, Sfan00 IMG, Shadowfyren, Shadowjams, Shakumafu, Shandris, Shanes, Shashank Shekhar, Shelmac, Sherylanne, Shibboleth, Shinjiman, Shnout, Sigma 7, Sigmundpetersen, Sikon, Silvergoat, SimonEast, SimonP, SimonTrew, Simple Bob, Simxp, SirGrant, Siroxo, Sjakkalle, Sjorford, Skirks, Skiwi, Skyeap, Slashuer, Slathering, Sleepeeg3, Slipknotmetal, SmackEater, Smokizzy, SmoothNikola, Smurfy, Smyth, Snaxe920, Snoyes, SoSaysChappy, Solphusion, Somaditya, Somebody in the WWW, Soumyasch, SouthernNights, Southpark20, Spe88, Spear of fire, SpecMode, Specter01010, SpeedyGonsales, Spencerperry, Sperling, Spliced, Spookfish, Spug, Srnelson, Ssd, Starionwolf, Steel, Steezly, StephenH, Stephenchou0722, Steve2011, Stevenh123, Stevenmitchell, Stevietheman, Stilroc, Stino v, Stonda, StuIsCool, StuThomas, SudoGhost, Supercooljs2, Supers, Supremeknowledge, Suruena, SusanLesch, Susvolans, Swabjob, Sweetback, Swollib, Swotboy2000, SyntaxError55, Syrthiss, Szhang21, TAG.Odessa, TJ Spyke, Ta bu shi da yu, Tacvek, Tadas12, TakuyaMurata, Tangotango, Tannin, Tarashav, Tawker, Tbhotch, Teamcritical, Tech2blog, Techman224, Techmdrn, Template namespace initialisation script, Tepidpond, Terence, Tfgbd, Thaek, Thalakan, Thavian, The Colclough, The Disco Times, The Epopt, The Fish, The Geneticist, The Rambling Man, The wub, TheChrisD, TheDoober, TheGeneralUser, TheKMan, TheNewPhobia, ThePointblank, Thealexweb, Theazman1, Thegreenj, Thetehror, Thewallowmaker, Thewikipedian, Thingg, Think outside the box, This, that and the other, Thisisntfake1, Thomas Larsen, Thompson.matthew, Thorpe, Throup, Thu, Thumperward, TiCPU, Tiggerjay, TimR, Timberframe, Timmeh, Timwi, Tinton5, ToasterOS, Tom Morris, Tomcage9, Tommy Irianto, Tompagenet, Tomy9510, Tonsofpcs, Tony1, Tothwolf, Towel401, Towsonu2003, Toyotaboy95, Tracer9999, Train2104, Treekids, Tregoweth, TrekMaster, Trevor MacInnis, Tripacer99, Trivialist, Trusilver, Tsunaminoai, Ttwaring, TubularWorld, TuneyLoon, Turionaltec, TuukkaH, Twernt, Twickline, TyA, Tyomitch, Tyraios, Tyrol5, UMC2, Uli, Ultimus, Umofomia, Unimaginative Username, Unixman83, Unknownperson1234, Uriel8, Utcursch, Vahid83, Valenciano, Vanished 6551232, Vanished user 5zariu3jisj0j4irj, Vbrtrmn, Vedek Dukat, Vegaswikian, Vincent.premysler, Vontafeijos, Vrenator, W1 m2, Wackymacs, Walter Görlitz, WalterGR, Wanderson9, Wangmike, Wapcaplet, Warlordwolf, Warren, Weararedjacket, Wengier, Werdan7, Werideatdusk33, Westley Turner, Whiteford8, Whkoh, Widefox, Wiki alf, Wiki fanatic, WikiBone, WikiFew, WikiMan44, Wikidexel2, Wikinger, WikipedianMarlith, Wikiwaka101, Windowsknowitall@msn.com, WindowzRULZlolZ, Windsok, Wknight94, Wlklpedla is meant to be vanda1ized, Wmahan, Wmplayer, Woohookitty, Wow1000, Wragge, Wsoder, Wykis, X1987x, XJamRastafire, Xation, Xelgen, Xeysz, Xgmx, Ximian99, Xlation, Xmachina, Xp54321, Xpclient, XrXeJoeXaXpXeXr, Xtreme racer, Xxglennxx, YUL89YYZ, Yama, Yamamoto Ichiro, Yamla, Yasimiazkhan, Yeeshenhao, Yehiahassan, Yergizmo, Ynhockey, Yousifnet, Yuckfoo, Yuhong, Z.E.R.O., Z98, ZFU738, Zakfleming, Zapvet, Zeeboid, Zepheus, Zer0431, Zntrip, Zondor, Zoney, Zundark, , , , 1542 anonymous edits

UDP_Data_Transport *Source*: http://en.wikipedia.org/w/index.php?title=UDP_Data_Transport *Contributors*: Attilios, Cburnett, Dawynn, E090, GTBacchus, Speciate, TReidLewis, 6 anonymous edits

User_Datagram_Protocol *Source*: http://en.wikipedia.org/w/index.php?title=User_Datagram_Protocol *Contributors*: 1exec1, 28421u2232nfenfcenc, A5b, Aapo Laitinen, Alfio, Ali@gwc.org.uk, Alison22, Alvin-cs, Aram33, Ardonik, Ashmoo, Ayeroxor, BAxelrod, Beacon11, Beefman, BenAveling, Bender235, Benqmonitor, Bentogoa, Bobby Caudwell, Bobstopper, Borgx, Bryan Derksen, Burisch, Cassan, Cburnett, Chenzw, Chiefmanzzz, Cincaipatrin, Conrad.Irwin, Conversion script, CraSH, CrookedAsterisk, Cvalda, DSatz, Damian Yerrick, Daniel Staal, DavidSol, DeadEyeArrow, DennyColt, DesTroy, Dfurbeck, Dicentra, Djordjes, Dkogh, E Wing, E090, Eagleal, Eclipsed, Enjoi4586, Erik Garrison, Esmond.pitt, Falcon9x5, Fennec, Fightin' Phillie, Flewis, Fredrikh, FrisoB, Frodet, Fromageestciel, Fubar Obfusco, Fudgefr, Gamera2, Giftlite, Ginkgo100, Gnipahellir, Golwengaud, Gordoni, Gorryf, Gracefool, Graham87, Gsmgm, Guy Harris, Hdante, Headbangerkenny, Hitherebrian, I dream of horses, IRedRat, Icestorm815, Ideoplex, InverseHypercube, Irrypride, Itai, JNighthawk, JTN, Jengelh, Jjinno, Jncraton, Joachim Wuttke, Joanjoc, Johnny 0, Johnuniq, Jtk, Jóna Þórunn, Kbrose, Kgfleischmann, Kinema, Kovianyo, Krellis, Kvng, Kwamikagami, Kyng, Lissajous, Ll1324, Locos epraix, MLeb, Mahjongg, Mam711, Mange01, Mani1, Materialscientist, Matt Darby, Michael Zimmermann, Mike Dill, Mike6271, Misterdom, Modulatum, Mpeg4codec, Mwholt, Neilc, Nixdorf, Nroets, Nubiatech, Od Mishehu, Ouimetch, Oxymoron83, PatheticCopyEditor, Paulish, PedroPVZ, Pelesl, Petri Krohn, Phatom87, Plindemann, Plugwash, Public Menace, Quiddity, R2richar, Razorflame, Recurring dreams, RedWolf, Rememberway, Reub2000, RevRagnarok, Reywas92, Rick Sidwell, Robbe, RobertL30, Romanm, Ronz, Rwwww, S-n-ushakov, SWAdair, Samersarhan83, Sammydee, Samuel, Sci13960, Smeggysmeg, Spearsall, Ssd293, Suruena, Tdangkhoa, Teapeat, Teemuk, Teh roflmaoer, Template namespace initialisation script, The Anome, The Monster, The Utahraptor, The-tenth-zdog, Thray, Tjdw, Tobias Bergemann, Towel401, Tpbradbury, Twinxor, UncleBubba, Upadhyaykc, Vanished user 5zariu3jisj0j4irj, Vinu Padmanabhan, Voidxor, WestwoodMatt, Wiarthurhu, Wkcheang, Xezbeth, Zara1709, ^demon, Александър, Тиверополник, 338 anonymous edits

Active_Directory *Source*: http://en.wikipedia.org/w/index.php?title=Active_Directory *Contributors*: 1exec1, 2mcm, 5 albert square, 7, AThing, Acdx, Achilees09, Adrian, Alereon, AlistairMcMillan, Alok.vij, Alro, Amniarix, Anaxial, Andre Engels, Andygreeny11, Anteaus, Arjun01, Art LaPella, Atomician, Av99, Avidmosh, Badgernet, Bairamkhan, Banano818, Barrylb, Ben Ben, Bishtmech, Bobprime, Bookgrrl, Bookofjude, Borgx, BrianWren, C45207, Canadian-Bacon, Canderson7, CanisRufus, Capricorn42, Carlosayam, Cbustapeck, Chasingsol, Chopeen, Chuunen Baka, Chzwhl, Chzz, Conversion script, Coremayo, Courcelles, Crkey, DMahalko, DRE, Dan East, DanShearer, DeepNorth, Defkkon, Defufna, Demiurge1000, Demonkoryu, Denisarona, DerHexer, Dhana82, Difu Wu, Disavian, Dlrohrer2003, Dogcow, DougsTech, Dpv, Dqd, DragonHawk, Drmies, Dysprosia, E. Ripley, E. Sn0 =31337=, EEMIV, EagleOne, Eagleridge, Ed!, EdBever, Edupedro, Edward, Emendation, EnamelWildcat, Extra999, Falcon8765, Favonian, Ff1959, Finlay McWalter, FleetCommand, Frap, Frencheigh, Gaff, GandalfDaGraay, Gary King, GcSwRhIc, Gdarin, Gkf101z, Gladmax, Gohiking, Groink, Gökhan, Hephaestos, Here, Highlandsun, Huffers, Igoldste, Imamusic, Init, JEETENDRAKAPSE, James086, Jann.poppinga, Jared Preston, Jasper Deng, Jay, Jeronim, Jesse Viviano, JesseGarrett, Jlundell, Joe.kraska, Joe407, Joelster, John Vandenberg, JohnWhitlock, Jonabbey, Jonathonblocker, Julesd, Justinm99, Jwreigh, Jxan3000, KD5TVI, Kasperd, KathrynLybarger, Kevin B12, Kingpin13, Kitamozihr, Kl4m, Kozuch, Kramsallad, Kungming2, Kwamikagami, Kycook, Kyle1278, Lankiveil, Lfstevens, Ligulem, Lilac Soul, Limbo socrates, Little Mountain 5, Lotje, Luvcraft, Luís Felipe Braga, Magnus.de, Marc Kupper, MarkWahl, MarsRover, Matlou, MaverickSolutions, Merlion444, Michael B. Trausch, Michael Hardy, Michael McDonnell, Michaelsoft, MidgleyDJ, Mikemaccana, Millerj90, Minna Sora no Shita, Mkdw, Moondyne, Moskvax, Mr little irish, MrOllie, Mrichards0, Muhandes, Muhgcee, Mullachv, Natalie Erin, Nigelm, Nikhilgulhane, Niklarin, Njan, Noahspurrier, Northamerica1000, Nullie, Ohconfucius, Omicronpersei8, Opes, Orfen, PaulGarner, Paulprout, Peter bertok, Pgan002, Phatom87, Philip Trueman, Pill, Pinkadelica, Plugwash, Pnm, Poccil, PoliticalJunkie, Prolog, Psy guy, Public Menace, PuerExMachina, QTCaptain, Qvestchun, Qwirty, Rcsheets, Rdeluca, Redvers, Reedy, Requestion, Retodon8, RexNL, Rich Farmbrough, Ricky81682, Rillian, Rjwilmsi, Rlian, Robinvasan, Roblu, Rosiehausler, Rshilwant, Rupert Clayton, Rusty Russell, RxS, Ryan Norton, RyanG, SJP, Sahil 11186, Salmon92, Sam Hocevar, SamBBaran, Saruman7, SaulPerdomo, Scot72, Script1, Senator Palpatine, Sendahmed, Shadowjams, Shane grimm, Sherly255, Sietse Snel, Sillygates, Singhsro, SixSix, Sniffnoy, Soifranc, Soumyasch, SpaceFlight89, Spearhead, Spidern, SpuriousQ, Srice13, StopBob, Sujay.shinde, Superm401, Suprax, Susfele, Susieber, Sveno, Syp, Ta bu shi da yu, Tactical, Tawker, Tbhotch, TehRabbitt, Tellyaddict, Terence, TestPilot, Tfl, The Thing That Should Not Be, Thecheesykid, Thedarxide, Thenoblenew, ThreeBlindMice, Tide rolls, Tim.alsop, Tobias Bergemann, Tom k&e, TomRawlinson, Totty3478, Tpk5010, Tristan Horn, Turb0chrg, TuxyQ, Twimoki, Twsx, Tyler.szabo, UltraViolet-64, Uncle G, Unknown entity, Vegaswikian, Vladislav Artukov, Volt42, W00tfest99, Warren, Welsh, Whisky drinker, Wiki alf, Wikijeff, Wikipelli, Windofkeltia, Wk muriithi, Wrs1864, Xenophonf, Xpclient, Yggdrasil42, Ykhwong, Yooden, Zachlipton, Zaxxon, Zidonuke, Zophar1, 975 anonymous edits

Image Sources, Licenses and Contributors

File:GroupLogic Logo.png *Source*: http://en.wikipedia.org/w/index.php?title=File:GroupLogic_Logo.png *License*: unknown *Contributors*: January, Kimsmith36, Tony Wills

File:Macintosh 128k transparency.png *Source*: http://en.wikipedia.org/w/index.php?title=File:Macintosh_128k_transparency.png *License*: unknown *Contributors*: Admrboltz, Avian, Brossow, Common Good, Grm wnr, Herbythyme, Himasaram, Kenmayer, Kevin chen2003, Mono, Pharos, 4 anonymous edits

File:Imac 16-9.png *Source*: http://en.wikipedia.org/w/index.php?title=File:Imac_16-9.png *License*: unknown *Contributors*: + Background : , under CC-by 3.0

File:Loudspeaker.svg *Source*: http://en.wikipedia.org/w/index.php?title=File:Loudspeaker.svg *License*: unknown *Contributors*: Bayo, Gmaxwell, Husky, Iamunknown, Mirithing, Myself488, Nethac DIU, Omegatron, Rocket000, The Evil IP address, Wouterhagens, 18 anonymous edits

File:Apple Macintosh Desktop.png *Source*: http://en.wikipedia.org/w/index.php?title=File:Apple_Macintosh_Desktop.png *License*: unknown *Contributors*: A104375, AlistairMcMillan, BorgQueen, Chmod007, Ctachme, Diego Moya, EdC, Ferrenrock, Grm wnr, HereToHelp, N. Harmonik, Ricardo Cancho Niemietz, 4 anonymous edits

File:Ad apple 1984.jpg *Source*: http://en.wikipedia.org/w/index.php?title=File:Ad_apple_1984.jpg *License*: unknown *Contributors*: Classicfilms, Edderso, Geniac, Grm wnr, HereToHelp, Ilse@, Jamoche, Lars T., Melesse, Mendaliv, Psantora, Rimshot, Zanimum, 5 anonymous edits

File:Apple-Macintosh.jpg *Source*: http://en.wikipedia.org/w/index.php?title=File:Apple-Macintosh.jpg *License*: unknown *Contributors*: User:Tmarki

File:MacII.jpg *Source*: http://en.wikipedia.org/w/index.php?title=File:MacII.jpg *License*: unknown *Contributors*: Grm wnr, Joonasl

File:Macintosh portable.jpg *Source*: http://en.wikipedia.org/w/index.php?title=File:Macintosh_portable.jpg *License*: unknown *Contributors*: Beavis, Grm wnr, J o, Kozuch, Kyro, Ranveig, Wpedzich, Wutsje, 1 anonymous edits

File:Macintosh classic.jpg *Source*: http://en.wikipedia.org/w/index.php?title=File:Macintosh_classic.jpg *License*: unknown *Contributors*: Grm wnr, Kilom691, Tipiac

File:Powerbook 100 pose.jpg *Source*: http://en.wikipedia.org/w/index.php?title=File:Powerbook_100_pose.jpg *License*: unknown *Contributors*: User:Danamania

File:Macintosh System 7.5.3 screenshot.png *Source*: http://en.wikipedia.org/w/index.php?title=File:Macintosh_System_7.5.3_screenshot.png *License*: unknown *Contributors*: Diego Moya, McLoaf, N. Harmonik, Remember the dot, 1 anonymous edits

File:MacOS8.6.jpg *Source*: http://en.wikipedia.org/w/index.php?title=File:MacOS8.6.jpg *License*: unknown *Contributors*: Calmer Waters, Fetchcomms, Fourthords, HereToHelp, Mono, N. Harmonik, Ozguy89

File:IMac Bondi Blue.jpg *Source*: http://en.wikipedia.org/w/index.php?title=File:IMac_Bondi_Blue.jpg *License*: unknown *Contributors*: Apalsola, Dodo, Grm wnr, HereToHelp, Ilse@, Thuresson

File:MacBook Pro situated on a wooden table.jpg *Source*: http://en.wikipedia.org/w/index.php?title=File:MacBook_Pro_situated_on_a_wooden_table.jpg *License*: unknown *Contributors*: Jeff Geerling

File:Mac mini.png *Source*: http://en.wikipedia.org/w/index.php?title=File:Mac_mini.png *License*: unknown *Contributors*: User:Wiki637

File:Macpro BW.png *Source*: http://en.wikipedia.org/w/index.php?title=File:Macpro_BW.png *License*: unknown *Contributors*: Bayo, Bukk, Kyro, Max Naylor

File:MacBook Air.png *Source*: http://en.wikipedia.org/w/index.php?title=File:MacBook_Air.png *License*: unknown *Contributors*:

Image:MacBook Pros.jpg *Source*: http://en.wikipedia.org/w/index.php?title=File:MacBook_Pros.jpg *License*: unknown *Contributors*: Benjamin Nagel

File:IMacG5guts.png *Source*: http://en.wikipedia.org/w/index.php?title=File:IMacG5guts.png *License*: unknown *Contributors*: Grm wnr

File:Windows logo.svg *Source*: http://en.wikipedia.org/w/index.php?title=File:Windows_logo.svg *License*: unknown *Contributors*: Blubberboy92, Cflm001, Colejohnson66, FleetCommand, JCRules, Johnuniq, Koman90, Mocha2007, Sudha Jagannathan, Train2104, Tyw7, Zzyzx11, 4 anonymous edits

File:Windows 7.png *Source*: http://en.wikipedia.org/w/index.php?title=File:Windows_7.png *License*: unknown *Contributors*: Addihockey10, Althepal, AnOddName, Anakin101, Andyso, AussieLegend, Bkell, Crazlunatic, Drilnoth, Feinoha, FleetCommand, GSK, Grayshi, Jan Hofmann, JetBlast, Jjupiter100, Josh the Nerd, Koman90, LOL, LobStoR, Mephiles602, Ngyikp, Nicholas Love, OriginalGamer, PhilKnight, RegularBreaker, Roscelese, S0aasdf2sf, SF007, SchuminWeb, Sdrtirs, Seaphoto, Sonicdude558, Sotcr, SpaceFlight89, The 888th Avatar, Warren, Wtshymanski, 35 anonymous edits

File:Windows 3.0 logo.svg *Source*: http://en.wikipedia.org/w/index.php?title=File:Windows_3.0_logo.svg *License*: unknown *Contributors*: Koman90, Sfan00 IMG, Skier Dude, Svgalbertian

File:Windows1.0.png *Source*: http://en.wikipedia.org/w/index.php?title=File:Windows1.0.png *License*: unknown *Contributors*: Aiyizo, Diego Moya, FleetCommand, Frogger3140, Gan Luo, Ghettoblaster, Happy Dude, Kubek15, Michaelkourlas, Neurolysis, Nicholas Love, Remember the dot, Warren, 11 anonymous edits

File:Windows 3.0 workspace.png *Source*: http://en.wikipedia.org/w/index.php?title=File:Windows_3.0_workspace.png *License*: unknown *Contributors*: Chugun, Courcelles, Dancraggs, FSII, FleetCommand, Nicholas Love, Shlomital, Tyomitch, Warren, Yamla, 5 anonymous edits

File:Windows 95 Desktop screenshot.png *Source*: http://en.wikipedia.org/w/index.php?title=File:Windows_95_Desktop_screenshot.png *License*: unknown *Contributors*: AlistairMcMillan, Caypartis, Damian Yerrick, Dan6hell66, Diego Moya, FleetCommand, Ghettoblaster, Koman90, Lordalpha1, McLoaf, Nicholas Love, Shlomital, ViperSnake151, Warren, 7 anonymous edits

File:WindowsCE7.png *Source*: http://en.wikipedia.org/w/index.php?title=File:WindowsCE7.png *License*: unknown *Contributors*: Koman90

File:Windows 8 Consumer Preview Start Screen.png *Source*: http://en.wikipedia.org/w/index.php?title=File:Windows_8_Consumer_Preview_Start_Screen.png *License*: unknown *Contributors*: User:Lifemaestro

File:Windows To Go USB Drive.png *Source*: http://en.wikipedia.org/w/index.php?title=File:Windows_To_Go_USB_Drive.png *License*: unknown *Contributors*: Adrignola, SF007, 2 anonymous edits

File:Windows Family Tree.svg *Source*: http://en.wikipedia.org/w/index.php?title=File:Windows_Family_Tree.svg *License*: unknown *Contributors*: J.int, Linfocito B, NOKIA 3120 classic, 1 anonymous edits

File:A help portal.gif *Source*: http://en.wikipedia.org/w/index.php?title=File:A_help_portal.gif *License*: unknown *Contributors*: Zureks

File:Icons-mini-page tree.gif *Source*: http://en.wikipedia.org/w/index.php?title=File:Icons-mini-page_tree.gif *License*: unknown *Contributors*: Avatar, Bayo, Rocket000

File:Icons-mini-page url.gif *Source*: http://en.wikipedia.org/w/index.php?title=File:Icons-mini-page_url.gif *License*: unknown *Contributors*: Avatar

File:Icons-mini-folder.gif *Source*: http://en.wikipedia.org/w/index.php?title=File:Icons-mini-folder.gif *License*: unknown *Contributors*: Avatar

File:Icons-mini-icon user.gif *Source*: http://en.wikipedia.org/w/index.php?title=File:Icons-mini-icon_user.gif *License*: unknown *Contributors*: Avatar, Rocket000

File:Active directory - OUs can not be given rights to objects.png *Source*: http://en.wikipedia.org/w/index.php?title=File:Active_directory_-_OUs_can_not_be_given_rights_to_objects.png *License*: unknown *Contributors*: DMahalko

MIX
Papier aus verantwortungsvollen Quellen
Paper from responsible sources
FSC® C105338

Printed by Books on Demand GmbH, Norderstedt / Germany